Handbook of Industrial Wastes Pretreatment

Handbook of Industrial Wastes Pretreatment

Jon C. Dyer
Arnold S. Vernick
Howard D. Feiler

WATER MANAGEMENT SERIES

Garland STPM Press
New York & London

15 14 13 12 11 10 9 8 7 6 5 4 3 2 1

Library of Congress Cataloging in Publication Data
Dyer, Jon C.
    Handbook of industrial wastes pretreatment.

    (Water management series)
    Includes index.
    1. Factory and trade waste.   2. Sewage—Purification.
3. Factory and trade waste—Law and Legislation—
United States.   I. Vernick, Arnold S., joint author.
II. Feiler, Howard, joint author.   III. Title.   IV. Series.
TD897.5.D93   628′.34   79-25702
ISBN 0-8240-7066-6

Published by Garland STPM Press
136 Madison Avenue, New York, New York 10016

Printed in the United States of America

Water Management Series

The Garland series in water management addresses the many aspects of water resource management which must be carefully integrated to protect the quality and quantity of our water supplies. The series is a comprehensive compilation of volumes dealing with the most important aspects of water resource management. Topics span the range from detailed information on design, cost, and performance of specific types of treatment processes to the environmental implications associated with certain management approaches. The series is designed to be a timely and authoritative reference source for practicing engineers and planners, as well as for regulatory agencies and students.

Series Editor:
GORDON L. CULP

Contents

1 / General Pretreatment Regulation (40 CFR, Part 403)   1

2 / Developing a Local Pretreatment Program   15

3 / Elements of a Pretreatment Program   27

4 / Impact of Industrial Wastes On Municipal Wastewater Treatment   147

5 / Sludge Disposal Considerations   191

6 / Industry's Role in Pretreatment   201

7 / Case Histories of Industrial Waste Control Programs   241

Preface

The control of industrial discharges into municipal wastewater treatment facilities is now a reality through the enactment of the Clean Water Act of 1977 (Public Law 95-217) and the promulgation of the General Pretreatment Regulation (40 CFR, part 403) on June 26, 1978. The thrust of the pretreatment regulation is to control indirect industrial dischargers and place the financial responsibility of eliminating toxic and priority pollutants from the environment where it should be—on the generators of the pollutants.

The success of the pretreatment effort and its subsequent benefit to the environment is directly associated with the cooperation and dedication of all groups involved; municipalities, industries, state and federal agencies, and the public. The development of mutual trust and cooperation is essential if the pretreatment of industrial wastes is to be realized.

The authors of this book have been fortunate to be involved with the USEPA in the pretreatment area and have strived to make this book and the material and information contained as up-to-date and applicable as possible. The authors want to thank the USEPA personnel that have been involved in the pretreatment program for all their assistance, efforts, and dedication. Appropriate references are included for pursuit of more detail if desired.

The intent of this book is to provide information on the development of local pretreatment programs and encourage the cooperative effort required between the local municipality and its respective industries.

Jon C. Dyer
April, 1980

List of Tables

List of Figures

1

General Pretreatment Regulation (40 CFR, PART 403)

INTRODUCTION

Historically, water pollution control efforts have utilized biological processes for the treatment of domestic wastes. Yet rapid growth of the industrial sector of our nation has introduced substances to our wastewater treatment works which cannot always be adequately treated by these conventional processes. Among these hard-to-treat products of industry are those which we call toxic and hazardous substances. (The interference with and inhibition of biological treatment processes by these substances are discussed in detail in Chapter 4.)

A recognized approach for the control of toxic and hazardous pollutants began in 1906 with the enactment of the Food and Drug Act and the Meat Inspection Act, which sought to prevent the distribution of consumer products containing injurious food preservatives (1). Although more advanced control measures were taken during the mid-1900s, the most progressive period has been the 1970s, with the creation of the United States Environmental Protection Agency (USEPA, or simply EPA). At present, mechanisms for the control of toxic pollutants are contained in federal legislation such as the Toxic Substances Control Act (PL 94-469), Resource Conservation and Recovery Act (PL 94-580), Clean Water Act (PL 95-217), Clean Air Act (PL 95-95), and the Safe Drinking Water Act of 1974 (PL 93-523), among others.

Recognizing the difficulties encountered with the introduction of industrial discharges containing toxic and hazardous wastes to municipal wastewater treatment plants, Congress enacted a series of amendments to the Federal Water Pollution Control Act (known as the Clean Water Act

1

after 1977) which specifically address this concern. These 1977 amendments required USEPA to develop pretreatment guidelines and standards which provide a uniform approach to the control of industrial pollutants introduced into publicly owned treatment works (POTWs). Pursuant to the mandate of Congress, USEPA has developed general pretreatment regulations for existing and new sources of pollution (Title 40 of the Code of Federal Regulations, Part 403). The objectives of the national pretreatment program are:

1. To prevent the introduction of pollutants into the POTW which will interfere with treatment operations and/or the use or disposal of the municipal sludge;
2. To prevent the introduction of pollutants into the POTW which will pass through the treatment works or otherwise be incompatible with the treatment works;
3. To correct inadequate treatment of many pollutants by industry and by POTWs prior to their release into the environment;
4. To improve the feasibility of recycling and of reclaiming the municipal and industrial wastewaters and sludges;
5. Generally, to reduce health and environmental risk of pollution caused by discharges to POTWs.

Figure 1–1 visually illustrates the environmental conditions impacted by the pretreatment program.

This text is intended to provide the foundation, impetus, and guidance required for the success of the national pretreatment program, not only from the federal standpoint, but also as related to local conditions, including the industry/municipal government interface. Critical to this approach is the need for a cooperative relationship between the POTW, industry, federal, state, and local government and the public.

BACKGROUND

The Federal Water Pollution Control Act of 1948 (FWPCA) represents the first national water quality legislation establishing a strategy for controlling water pollution. This strategy was based on cleanup requirements for the desired uses of effluent-receiving waters (drinking water, body contact, recreation, fishing, etc.) as determined by State governments and the water quality conditions necessary to support those uses. This strategy was generally ineffective because of a number of political, technical, and legal weaknesses:

1. Stream use designations tailored to attract or discourage industrial development;

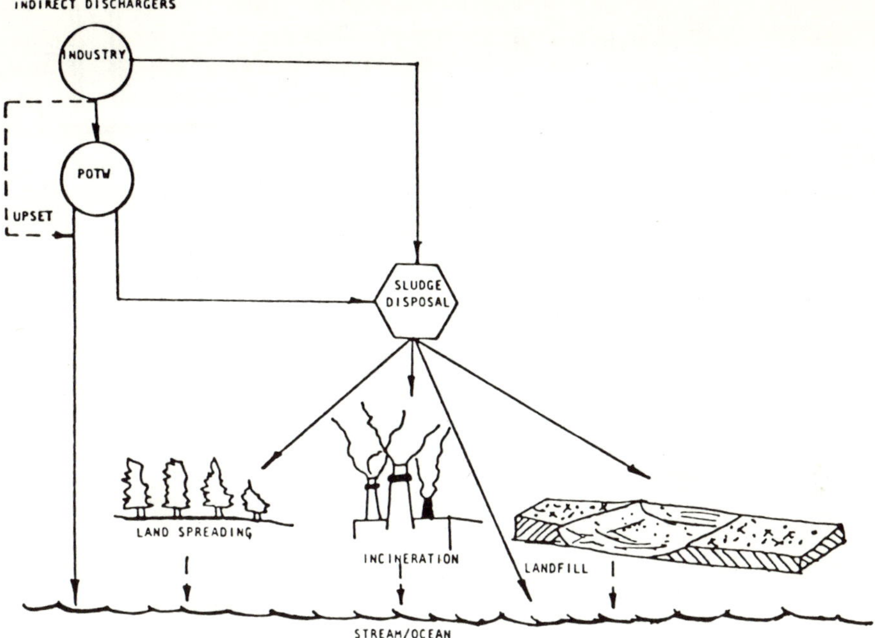

Figure 1-1 / Overview of pretreatment environmental problem (2).

2. Inadequate information on the cause-and-effect links between discharges and water quality;
3. Lack of national uniformity of standards and in enforcement;
4. Problems of equity between old and new pollution sources;
5. General lack of attention to pollution emanating from sources other than a pipe.

The FWPCA underwent substantial amendment in 1972 by passage of Public Law 92-500. In this form, the act accomplishes three basic tasks:

1. Regulation of discharges of pollutants from point sources, primarily industrial plants, municipal sewage treatment plants, and agricultural feed lots;
2. Regulation of spills of oil and hazardous substances;
3. Financial assistance for sewage treatment plant construction.

With the 1972 amendments, the FWPCA established separate regulatory schemes for two classes of point source dischargers: persons discharging directly into navigable waters and persons discharging into publicly owned treatment works (indirect dischargers) (3). Figure 1-2 depicts this division for industrial dischargers.

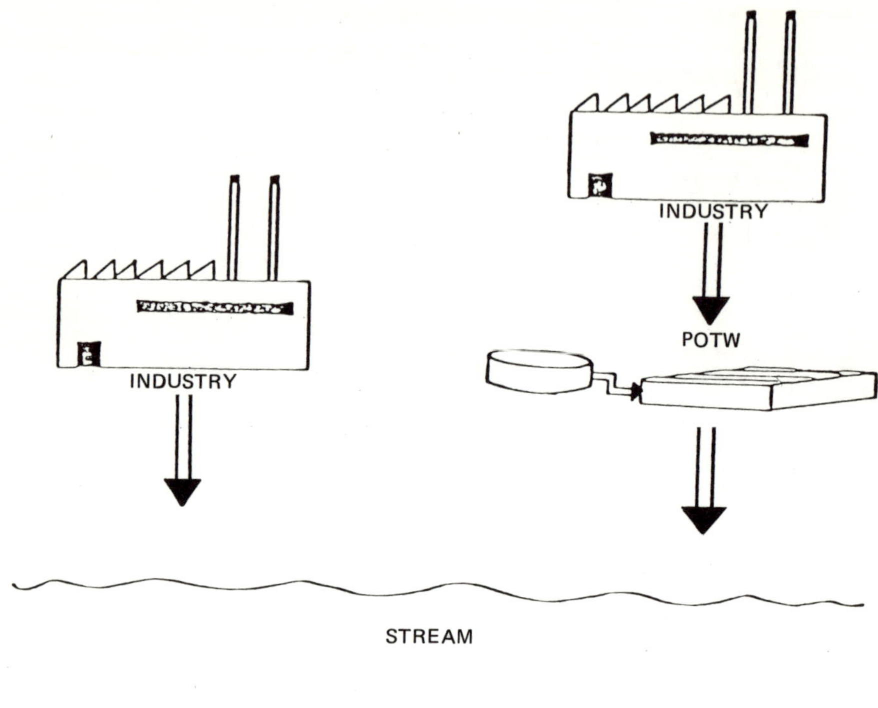

Figure 1-2 / Direct/indirect industrial discharger (2).

The direct dischargers are then subject to a dual set of requirements: effluent standards and water quality standards. Effluent standards are limitations on the amounts of pollutants that may be discharged by particular types of dischargers; they are distinct from water quality standards, which consist of rules defining requirements for the quality of ambient water. Effluent standards (limitations) are to be established under best-available-technology requirements.

Under section 307 of the FWPCA, dischargers into POTWs may come under regulation. This section of the act requires EPA to promulgate pretreatment standards designed to prevent the discharge of pollutants through POTWs which are determined not to be susceptible to treatment by POTWs, or which would interfere with the operation of POTWs.

Even with the passage of PL 92-500, Congress clearly anticipated that review, course corrections, and fine tuning of the act would soon be required. In 1976, EPA entered into a consent decree in the case of NRDC v. Train with the Natural Resources Defense Council, Inc. (NRDC); Environmental Defense Fund, Inc.; Businessmen for the Public Interest, Inc.;

National Audubon Society, Inc.; and Citizens for a Better Environment. The consent decree requires EPA to review 65 classes of toxic pollutants which may be discharged by 21 industrial categories and to establish pretreatment standards for any of the 65 classes of toxic pollutants which are found not to be susceptible to treatment by POTWs or would interfere with the operation of POTWs. On March 9, 1979, the NRDC consent decree specifying 21 industrial categories was expanded to encompass 34 industrial categories, as listed in Table 1-1 (NRDC v. Costle). The consent decree also requires that the pretreatment standards cover 95% of the industrial users in each industrial category.

The realization that the water quality field is changing rapidly (new problems emerge, new technologies develop, and overall knowledge improves apace) led to the Clean Water Act of 1977 (PL 95-217), which substantially revises the mandates contained in the FWPCA. The Clean Water Act (CWA) incorporated substantial portions of the NRDC agreement and brought about other changes in the direction of improved water quality. Probably the most important of the 1977 amendments are those which expressed concerns about chemical pollution, materials recycling, and environmentally compatible technical systems. It was realized that there are different forms of pollution and that some kinds pose a greater threat to public health and the environment than others. This recognition, combined with emerging questions about the cost effectiveness of applying stringent, technology-based limitations on "ordinary" wastes, led in 1977 to a new

Table 1-1 / NDRC v. Costle Consent Decree—34 Industries (4)

1. Adhesives	18. Pulp and paper
2. Leather tanning[a] and finishing	19. Textile mills[a]
3. Soaps and detergents	20. Timber[a]
4. Aluminum forming	21. Coal mining
5. Battery manufacturing	22. Ore mining
6. Coil coating	23. Petroleum refining[a]
7. Copper forming	24. Steam electric[a]
8. Electroplating[a,b]	25. Organic chemicals
9. Foundries	26. Pesticides
10. Iron and steel	27. Pharmaceuticals
11. Nonferrous metals[a]	28. Plastic and synthetic materials
12. Photographic supplies	29. Rubber
13. Plastics processing	30. Auto and other laundries
14. Porcelain enamel	31. Mechanical products
15. Gum and wood chemical	32. Electric and electronic components
16. Paint and ink	33. Explosives manufacturing
17. Printing and publishing	34. Inorganic chemicals[a]

[a]Industries for which interim pretreatment standards were promulgated in 1977 and 1978.

[b]While interim guidelines were developed for the electroplating industry, these were suspended on May 14, 1979. EPA has since promulgated the pretreatment standards for the electroplating point source category (September 7, 1979).

classification of pollutant types, with different requirements specified for each category. These changes result in a much greater emphasis on the control of toxic pollutants with both direct dischargers to navigable waters and dischargers to POTWs (indirect dischargers) equally sharing the responsibility for the burden of controlling point source pollution. The 1977 legislation also promotes recycling and reuse of pollution control byproducts (effluent, sludge, and nutrients), energy conservation, and multiple use of lands and waters which are components of wastewater treatment systems. Specific to the national pretreatment program, the modifications made in 1977 under section 307 of CWA demonstrate the congressional intent to reduce toxic pollution and to enhance the feasibility of recycling and reclaiming sludges and effluents generated by POTWs.

As required by section 307 of the FWPCA as amended in 1972, EPA developed pretreatment standards, Title 40 of the Code of Federal Regulations, Part 128, in November 1973. On February 2, 1977, EPA proposed a rule which would establish mechanisms and procedures for enforcing national pretreatment standards controlling the introduction of nondomestic wastes into POTWs.

OVERVIEW

Fundamental changes instituted by PL 92-500 and the CWA require POTWs to obtain permits for their discharges and meet minimum effluent standards. These actions by Congress have a direct bearing on the control of industrial pollutants introduced to these systems.

National Pollutant Discharge Elimination System

Procedures developed under section 402 of CWA provide details for implementation of the National Pollutant Discharge Elimination System (NPDES) permit program. Under the program, all point sources, including POTWs, must obtain a permit to discharge to navigable waters of the United States. NPDES permits are not required for industrial sources contributing to POTWs, but many limitations placed on a POTW, beyond standard secondary treatment, are aimed at controlling the effects of nondomestic wastewaters.

The NPDES permit program affects state and local pretreatment policies in two principal ways. First, as a part of the permit application, the permittee must develop and submit a plan to the state or local water quality planning agency for implementing a pretreatment program. Such a submission would include the following subdivisions:

1. *Legal authority:* The POTW must operate under legal authority which will enable it to apply and enforce the applicable sections of the Act and any regulations implementing those sections.
2. *Compliance procedures:* The POTW must develop procedures to ensure compliance with the POTW pretreatment program.
3. *Adequate resources:* The POTW must show sufficient resources (funds and personnel) to operate an effective program.

Secondly, NPDES permit regulations require the permittee to provide notice to the regional administrator of EPA of the following:

1. Any new introduction of pollutants into such treatment works from a source which would be a new source as defined in Section 306 of the Act if such source were discharging pollutants;
2. Any new introduction of pollutants which exceeds 10,000 gallons on any one day into such treatment works from a source which would be subject to section 301 of the Act if such source were discharging pollutants;
3. Any substantial change in volume or character of pollutants being introduced into such treatment works by a source introducing pollutants into such works at the time of issuance of the permit.

This notice must include information on the quantity and quality of wastewater introduced by the new source into the POTW, and any anticipated impact on the effluent discharged from the works.

National Standards of Performance

Since most POTWs discharge their effluents to navigable waters, the Act requires that effluent-limiting regulations be promulgated. EPA has established rules and regulations governing POTW discharges under the heading of Secondary Treatment Information. These regulations set forth specific concentration limits to be achieved by secondary wastewater treatment facilities. Limits are placed on permissible discharge concentrations (or removal efficiencies) for biochemical oxygen demand (BOD), suspended solids (SS), and fecal coliform bacteria. Additionally, an acceptable pH range for secondary POTW effluents is set. The pollutants limited in the regulation are generally susceptible to treatment in POTWs. As a consequence, industrial pollutants only become important with regard to secondary treatment standards when the contribution causes these pollutant discharge limitations to be exceeded.

In many instances, state or local water quality standards require a degree of treatment greater than that required to meet the secondary treatment regulations, such as more stringent BOD or SS requirements.

Additionally, water quality standards often exist for pollutants other than those regulated in the secondary treatment standards, such as phosphorus, nitrogen compounds, metals, etc. Treatment requirements necessary to meet water quality standards are normally incorporated into the NPDES permit. Consequently, control of applicable industrial pollutants can be extremely important in helping the municipality meet its NPDES permit requirements.

General Pretreatment Regulation

The national pretreatment strategy and the general pretreatment regulation (40 CFR, Part 403) provide for national pretreatment standards. These pretreatment standards will include general discharge prohibitions that apply to all users of a POTW who discharge nondomestic wastes, as well as standards applicable to specific industrial categories. The objectives behind the pretreatment program are:

1. To prevent the introduction of pollutants into the POTW which will interfere with treatment operations and/or the use or disposal of the municipal sludge;
2. To prevent the introduction of pollutants into the POTW which will pass through the treatment works or otherwise be incompatible with the treatment works;
3. To correct inadequate treatment of many pollutants by industry and by POTWs prior to their release into the environment;
4. To improve the feasibility of recycling and of reclaiming the municipal and industrial wastewaters and sludges;
5. Generally, to reduce health and environmental risk of pollution caused by discharges to POTWs.

Given the complexity of the problem and the goals to be obtained, EPA, through the general pretreatment regulation, is providing the framework for cleanup of industrial pollutant discharges to POTWs and designating responsibilities and deadlines for compliance to everyone involved including EPA, states, POTWs, industry, and the public.

This will ideally result in better coordination among municipalities and industry and provide a consistent and equitable approach toward nondomestic discharger's use of POTWs on a national level. The general pretreatment regulations are aimed at reducing pollutant discharges to a POTW which interfere with its operation, contaminate the effluent or sludges, or otherwise pass through the POTW inadequately treated to the environment. The effort required to achieve these objectives requires not only a commitment from industry and POTWs, but also the dedication of substantial resources as well as public and political support at the local, state, and national level.

The general pretreatment regulations (40 CFR, Part 403) implement many sections of the FWPCA as amended by the Clean Water Act of 1977. This regulation applies to:

1. Nondomestic pollutants discharged into or transported and introduced into POTWs;
2. POTWs which receive wastewaters from sources subject to national pretreatment standards;
3. States which have NPDES programs;
4. Any new or existing source subject to pretreatment standards.

To ensure that the goals of the national pretreatment policy are obtained, the general pretreatment regulations provide for the development of two types of national standards. Sections 307(b) and (c) of CWA provide the authority for developing these standards: prohibited discharge standards and categorical standards.

Prohibited Discharge Standards

These standards prohibit the discharge of any nondomestic wastes containing certain types or amounts of pollutants which would interfere with the operation of a POTW. This restriction applies to all users of a POTW and is designed to protect the mechanical and hydraulic integrity of POTWs. The following pollutants are considered prohibited discharges:

1. Substances which create a fire or explosive hazard;
2. Corrosive materials and discharges with a *pH* less than 5 (unless the POTW is specifically designed to accommodate corrosive materials);
3. Solid or viscous materials in amounts which obstruct flow or interfere with operation of a POTW;
4. Any pollutant (including BOD and SS) released in a discharge at a flow rate and/or pollutant concentration which a discharger knows or has reason to know will cause interference to the POTW;
5. Heat discharges which will inhibit biological activity in the POTW treatment plant resulting in interference or causing damage, or which will create temperatures which exceed 65°C (150°F) at the POTW and 40°C (104°F) at the POTW treatment plant unless the approval authority (EPA or NPDES state), upon request by the POTW, approves other temperature limits.

As mentioned previously, these prohibitions apply to all nondomestic discharges to POTWs, whether or not the user is subject to any other national or pretreatment requirements.

Any POTW developing a pretreatment program is required to develop and enforce specific limits for these pollutants based on its capacity to handle such pollutants as a requirement of the program. Where these pollutants are contributed by a user or users and subsequently cause or significantly contribute to a violation of a POTW's NPDES permit, and where the violation is likely to reoccur, the POTW should develop and enforce specific effluent limits for the responsible user or users. Such limits are to be incorporated into the POTW's NPDES permit. If no action has been taken by the POTW within 30 days after notice by EPA of the violation, EPA may take appropriate action.

POTWs are required to comply with the prohibited discharge standards beginning on the effective date of the general pretreatment regulations (August 25, 1978) except for the heat discharge limitations which must be complied with within three years after August 25, 1978 (40 CFR, part 403.5 (e)).

Categorical Standards

The categorical pretreatment standards specify quantities or concentrations of pollutants or pollutant properties which may be discharged or introduced to POTWs by new or existing sources in specific industrial subcategories. These standards are being published as separate regulations and revised under authority of sections 307(b) and (c) of CWA.

The NRDC consent decree and CWA required EPA to review 65 classes of toxic pollutants (Table 1–2) which may be discharged by the 34 industrial categories (Table 1–1) (5).

For existing users of a POTW, the standards would contain numerical pollutant discharge limitations for each industrial subcategory based upon the best available technology economically achievable (BATEA, or simply BAT) or any more stringent effluent standards under section 307(a) of CWA. New source standards will be based solely upon the BATEA in each industrial subcategory.

Although categorical standards will be normally expressed as concentration limits, whenever possible, equivalent mass limits will be provided so that the enforcement authorities may use either concentration or mass limits. Concentration limits in categorical pretreatment standards would apply to the effluent of the process being regulated by the standard, unless otherwise stated by the specific standard.

On June 26, 1978, the general pretreatment regulations previously discussed (40 CFR, Part 403) were promulgated and became effective August 25, 1978, replacing the existing pretreatment standards (40 CFR, Part 128). Proposed amendments to 40 CFR, Part 403 were published in the Federal Register October 29, 1979, and January 16, 1980.

SCOPE

The NRDC consent decree requires EPA to review 65 classes of toxic pollutants (Table 1–2) which may be discharged by the 34 industrial categories, and to establish pretreatment standards for any of the 65 classes of toxic pollutants and any other nontoxic pollutants which are found not to be susceptible to treatment by POTWs or would interfere with the operation of a POTW. The 65 pollutants were selected on the basis of a review of available data on 232 pollutants of greatest environmental concern. The 65 classes of toxic pollutants include: substances for which there is evidence of carcinogenicity, mutagenicity, and/or teratogenicity; and substances known to have toxic effects on human or aquatic organisms. EPA has concentrated its standard setting effort on the industries believed to be discharging the

Table 1-2 / 65 Classes of Toxic Pollutants Listed in Consent Decree and Referenced in 307(a) of the CWA of 1977

Acenapthene	Endosulfan and metabolites
Acrolein	Endrin and metabolites
Acrylonitrile	Ethylbenzene
Aldrin/Dieldrin	Fluoranthene
Antimony and compounds	Haloethers
Arsenic and compounds	Halomethanes
Asbestos	Heptachlor and metabolites
Benzene	Hexachlorobutadiene
Benzidine	Hexachlorocyclopentadiene
Beryllium and compounds	Hexachlorocyclohexane
Cadmium and compounds	Isophorone
Carbon tetrachloride	Lead and compounds
Chlordane	Mercury and compounds
Chlorinated benzenes	Napthalene
Chlorinated ethanes	Nickel and compounds
Chloralkyl ethers	Nitrobenzene
Chlorinated naphthalene	Nitrophenols
Chlorinated phenols	Nitrosamines
Chloroform	Pentachlorophenol
2-Chlorophenol	Phenol
Chromium and compounds	Phthalate esters
Copper and compounds	Polychlorinated biphenyls (PCBs)
Cyanides	Polynuclear aromatic hydrocarbons
DDT and metabolites	Selenium and compounds
Dichlorobenzenes	Silver and compounds
Dichlorobenzidine	2, 3, 7, 8, -Tetrachlorodibenzo-p-dioxin (TCDD)
Dichloroethylenes	Tetrachloroethylene
2, 4-dichlorophenol	Thallium and compounds
Dichloropropane and dichloropropene	Toluene
2, 4-Dimethylphenol	Toxaphene
Dinitrotoluene	Trichloroethylene
Diphenylhydrazine	Vinyl chloride
	Zinc and compounds

Table 1-3 / Number of Indirect Dischargers for the 34 Consent Decree Industries (2)

| | Indirect dischargers | |
Industry	Total	Total number to be regulated by pretreatment regulation
Auto and other laundries	90,000	1,500
Coal mining	0	0
Electroplating	6,568	6,568
Inorganic chemicals	120	108
Iron and steel	238	237
Leather	170	170
Machinery and mechanical products		
Mechanical products	71,000	3,000
Battery manufacturing	170	170
Plastics processing	2,700	1,350
Foundries	880	880
Coil coating	50	50
Porcelain enamel	138	138
Aluminum forming	70	70
Copper forming	57	57
Electric and electronic components	8,000	500
Photographic supplies	110	110
Miscellaneous Chemical Manufacturing		
Pesticides	46	37
Gum and wood chemicals	7	6
Pharmaceuticals	200	160
Explosives	0	0
Adhesives and sealants	500	500
Carbon black	1	1
Nonferrous metals	79	57
Ore mining	0	0
Organic chemicals	900	720
Paint and ink	1,600	460
Printing and publishing	63,000	2,300
Paving and roofing	138	0
Petroleum refining	48	48
Plastic and synthetic materials	160	128
Pulp and paper	260	50
Rubber	537	9
Soaps and detergents	450	400
Steam electric	150	150
Textile mills	1,700	915
Timber products	189	49
Total	250, 236	20, 898

NOTE: Numbers shown are best available estimates (1/15/79) based on best available information, and are probably maximum.

pollutants of greatest environmental concern, specifically the 34 industrial categories listed in the consent decree (Table 1-1). Eight of the industries identified are those for which interim pretreatment standards were promulgated in 1977 and 1978 pursuant to the consent decree.

Looking at current information from the first ten industrial categories being reviewed for BAT standards, EPA anticipates that many industries of the 34 major industrial categories (and some entire categories) will not require pretreatment standards because they may fall under an exclusion of the consent decree. Current EPA estimates indicate there are more than 20,000 industrial dischargers to POTWs in the 34 industrial categories to be considered initially in the focus for pretreatment standards (5). Table 1-3 illustrates the EPA estimate for the anticipated number of indirect dischargers to be regulated by the general pretreatment regulation. A large portion of these industries, potentially subject to national pretreatment standards, discharge to approximately 3,000 of the more than 13,000 POTWs in the United States. About 1,000 of the POTWs affected are providing levels of treatment greater than secondary treatment. The national pretreatment program is expected to require compliance with pretreatment standards by up to 20,000 dischargers in 1980, and potentially as many as 38,000–55,000 by 1983.

There are approximately 568 POTWs which are designed to accept wastewater flows of more than 5 mgd and which are believed to receive industrial wastes subject to national pretreatment standards (5). These 568 POTWs account for approximately 87% of the industrial influent to POTWs. EPA and states with approved NPDES programs will be responsible for enforcement of national pretreatment standards in the almost 1,900 POTWs receiving industrial wastewaters which are not required to develop pretreatment programs.

REFERENCES

1. Temple, Truman. Controlling toxics. *EPA Journal,* vol. 5, No. 7, pp. 12–15, July/August 1979.
2. *Local Pretreatment Program Requirements and Guidance.* Springfield, Virginia: Environmental Technology Consultants, Inc., June 1979.
3. Dolgin, E. L. and Guilbert, T. G. P., eds., *Federal Environmental Law.* St. Paul, Minnesota: West Publishing Co., 1974.
4. Natural Resources Defense Council, Inc., et al., Plaintiffs, v. Douglas M. Costle, Administrator of the Environmental Protection Agency, et al., Defendants, Nos. 2153-73, 75-0172, 75-1698, and 75-1267, March 9, 1979.
5. General pretreatment regulations for existing and new sources of pollution. *Federal Register,* vol. 43, no. 123, June 26, 1978.

Developing a Local Pretreatment Program

POTW COMPLIANCE

Primary responsibility for ensuring compliance with and enforcing the pretreatment standards will rest upon the local authorities responsible for the POTW pretreatment program (POTW control agency) (1). The POTW control agency will have the most immediate commitment to the success of the program, to proper operation of the POTW, to increased opportunity for sound sludge management methods, and to local protection of the environment and public health. This idea of local control is initiated and required by the Clean Water Act. Section 402(b)(8) of the Act requires the establishment of pretreatment programs enforceable through each POTW's NPDES permit. Also, Section 309(f) of the Act allows for civil action to be taken against a POTW for not enforcing the national pretreatment standards. POTWs which receive effluents from significant industrial sources subject to pretreatment standards, (prohibited discharge and categorical standards) must establish a POTW pretreatment program. The exception to this requirement is that small POTWs which would not have adequate resources to implement an effective program may not be required to develop a POTW pretreatment program.

Any POTW (or combination of POTWs operated by the same authority) with a total design flow greater than five (5.0) million gallons per day (MGD) and receiving wastes from sources subject to national pretreatment standards or pollutants which pass through untreated or interfere with the operation of the POTW are required to establish a POTW pretreatment program as a condition of their NPDES permit. Authorities of POTWs with total design flows of 5.0 MGD or less will be required to develop a pretreatment program when the following conditions hold (1):

1. They wish to modify the pollutant discharge limits in categorical standards to account for POTW removal of a pollutant.
2. The regional administrator of EPA or the director of an NPDES state finds that the nature or volume of the industrial influent upsets the treatment process, causes a violation of the POTW effluent limitations, or contaminates the municipal sludge, or when the director finds other circumstances which warrant the pretreatment program in order to prevent the passthrough of untreated pollutants or interference with the POTW.

The approval authority (either an NPDES state with an approved pretreatment program or EPA) will assume primary responsibility for enforcing pretreatment standards in the 5-MGD-or-smaller POTWs not having POTW pretreatment programs.

POTWs, whether or not they are federally funded, and new and existing users of POTWs are subject to the general pretreatment regulations (40 CFR, Part 403). POTWs which meet the requirements of this chapter must receive approval of a POTW pretreatment program within three years after reissuance or modification of its existing NPDES permit, but in no case later than July 1, 1983. The program requirements will be administered by the POTW control agency to assure compliance by industrial users with applicable pretreatment standards. The following procedures will be essential requirements to obtaining an approved program:

1. Evaluate and establish the proper legal authority;
2. Establish compliance procedures to enforce the pretreatment requirements;
3. Determine the level of resources required;
4. Conduct an industrial waste survey (IWS);
5. Develop an industrial waste ordinance;
6. If requested, determine the POTW removal allowance;
7. Take note of fundamentally-different-factors (FDF) determinations;
8. Develop an organizational structure;
9. Develop enforcement procedures to ensure compliance;
10. Submit a detailed description of the pretreatment program for NPDES state or EPA review and approval;
11. Take note of the state options for statewide control of local pretreatment programs.

Each of these requirements will be developed in detail in subsequent sections of this chapter.

POTW PRETREATMENT PROGRAM SUBMISSION

The contents of a POTW pretreatment program submission must contain suitable information, in sufficient detail, to adequately demonstrate the POTW's ability and plan of implementation to carry out the pretreatment program.

Prior to submission of three copies of the program description to the NPDES state approval authority (or to the regional administrator of EPA for POTWs in non-NPDES states), the POTW control agency should provide information to and consultation with interested and affected members of the public. A copy of the draft submission should be available to the public thirty days before it is submitted to the approval authority. This draft submission should also be accompanied by a fact sheet written in layman's terms, adequately describing the POTW pretreatment program and its significance and/or any request for authority to modify categorical pretreatment standards for pollutants removed by the POTW.

In order to provide for uniformity, consistency of reviews, and expediency in approvals, the following outline has been designed to aid in the submission of a POTW pretreatment program (2).

I. Program Description

A. Cover Letter

1. Statement from the city solicitor, a city official acting in a comparable capacity or an attorney for the POTW (for those which have independent legal counsel) that the POTW has authority adequate to carry out the pretreatment program described in 40 CFR, Part 403, section 403.8. This statement must:
   a. Identify the provision of legal authority under section 403.8 (f);
   b. Identify the manner in which the POTW will implement the program requirements set forth in section 403.8. This will include the POTW's procedures to ensure compliance with the requirements of the pretreatment program;
   c. Identify the means by which the pretreatment standards will be applied to industrial users (e.g., by order, permit, ordinance, contract, etc.);
   d. Identify how the POTW intends to ensure compliance with pretreatment standards and requirements, and in the event of noncompliance how the POTW intends to enforce the standards and requirements.

B. Supporting Statutes and Regulations

1. A copy of any statues, ordinances, regulations, contracts, agreements, or other authorities relied upon by the POTW for its administration of the program. A statement reflecting the endorsement or approval of the local boards or bodies responsible for supervising and/or funding the POTW pretreatment program shall also be included.

C. POTW Organization

1. A brief description of the POTW with organization charts. If more than one agency is responsible for administering the program, identification, respective responsibilities, and procedures should be explained.

D. Resources

1. A description of the funding levels and full- and part-time manpower available to implement the program.

II. Fact Sheet

A. A description of the POTW pretreatment program and its significance and/or any request for authority to modify categorical pretreatment standards for pollutants removed by the POTW. This fact sheet should be written in layman's terms and will accompany a copy of the draft submission for public review.

III. Summary of Public Participation

A. The submission should include:

1. A summary of public participation efforts;
2. Major public comments received in conjunction with adopting any statute, ordinance, contract, agreement or other authority for enforcing pretreatment standards;
3. The manner in which major issues identified by the public have been resolved.

IV. **Request for Authority to Revise Categorical Standards**

    A. Applications for authorization to revise categorical discharge limits may be included with the pretreatment program submission pursuant to Sections 403.8, 403.9 and 403.11 of the general pretreatment regulation.

V. **Request for Conditional Approval**

    A. The POTW may request conditional approval of the pretreatment program pending the acquisition of funding and personnel for certain elements of the program. This request for conditional approval must meet the requirements set forth above except that the submission must demonstrate that:

        1. A limited aspect of the program does not need to be implemented immediately;

        2. The POTW has adequate legal authority and procedures to carry out those aspects of the program which will not be implemented immediately;

        3. Funding and personnel for the program aspects to be implemented at a later date will be available when needed (the POTW will describe in the submission the mechanism by which this funding will be acquired).

To ensure that local pretreatment programs are established, the state or appropriate permit-issuance authority will develop compliance schedules as part of the NPDES permit.

MODIFICATION OF CATEGORICAL PRETREATMENT STANDARDS

Introduction

As discussed in Chapter 1, categorical pretreatment standards specify quantities or concentrations of pollutants or pollutant properties which may be introduced to a POTW by new or existing users in specific industrial subcategories. Should a POTW remove all or any part of a toxic pollutant subject to the categorical standards, then the pretreatment requirements for the source(s) actually discharging that pollutant may be revised to reflect the removal of the pollutant by the POTW. Section 307(b)(1) authorizes this revision of categorical pretreatment standards, provided that the discharge from the POTW does not violate that effluent limitation or standard which would be applicable to the pollutant if it were discharged other than through a POTW, and does not prevent sludge use or disposal in accordance with section 405 of the Clean Water Act.

Section 403.7 provides the procedures and requirements for modifying categorical pretreatment standards to reflect POTW removal. Revision of these standards shall be on a case-by-case basis. POTW removal of a specific pollutant means reduction or alteration of the nature of a pollutant in the influent by the POTW to a less toxic or harmless state prior to discharge to the receiving waters. This reduction or alteration may be accomplished by physical, chemical, or biological processes. This removal may be incidental to the operation of the POTW or specifically designed into the POTW. Removal allowance may be given for in-line treatment of biodegradable pollutants where removal can be calculated even if no detectable amount is found in the POTW influent. It should be made clear that dilution will not be considered in approving removal allowance. Also, the inability of monitoring equipment to analyze and detect diluted nondegradable pollutants in influents or effluents will not be considered at meeting removal criteria.

Requirements For Revision Of Categorical Pretreatment Standards

Introduction

Any POTW receiving wastes from sources who are or may in the future be subject to the categorical pretreatment standards may submit application for authorization to revise discharge limits from the regional administrator of EPA and/or the director of an NPDES state (i.e., from the appropriate approval authority).

The approval authority will receive, review, and evaluate the application to determine if the removal allowance requested is justified, and documentation provided is sufficient to warrant the revised limit. It should be noted that the value of modifying a standard may be questionable if, in fact, the subsequent cost to the industrial user for funding required sampling, analysis, documentation, etc., exceeds the reduction in the cost of technology installed at the industrial installation to meet the modified standard.

The approval authority will advise the POTW that a preliminary engineering review of the potential savings to industrial users versus the costs of documenting the removal allowance be made prior to embarking on the modification submission. The POTW should also be advised to consider the financial implications of the impact of the removal allowance on municipal sludge disposal or utilization practices and that all industrial users discharging a given pollutant must be considered in any determination.

The review of removal allowance submissions will focus on the existing performance record of the particular POTW and potential causes for any previous upsets of permit violations. An analysis of the potential capabilities of pollutant removal through the unit processes and flow train of the POTW will be made. An in-depth review of the performance data (solid/liquid) presented should be made. Some backup analytical support is helpful in determining the validity of the data presented. The potential impact of the removal allowance on the sludge management techniques used at the POTW should be considered.

Requirements

The following requirements must be met in order for a POTW to revise pollutant discharge limits in categorical pretreatment standards (1):

1. The POTW has a pretreatment program (approved in accordance with sections 403.8, 403.9, and 403.11). A POTW may conditionally revise the discharge limits for specific pollutants, even though a pretreatment program has not been approved, if this is in accordance with the terms and conditions of section 403.7(b)(2).
2. The POTW provides consistent removal of each pollutant for which the discharge limit in a categorical standard is to be revised, at a level which justifies the amount of revision to the discharge limit. (Consistent POTW removal is defined in section 403.7(a)(1).)
3. POTWs with combined sewers or bypass systems (which discharge at least once annually) may claim consistent removal of a pollutant only if: (a) the industrial user (IU) can provide containment or cease all discharges of the pollutant during circumstances in which a bypass event can be expected to occur, or (b) the consistent removal claimed is reduced according to the following equation:

$$Y = \frac{X}{1-r} \times \frac{8760-Z}{8760}$$

where:

X = pollutant discharge limit specified in the applicable categorical pretreatment standard (mg/l)

r = consistent POTW removal rate percentage expressed in decimal form

Y = revised discharge limit for the specified pollutant (mg/l)

Z = hours per year that bypassing occurred between the IU(s) and the POTW treatment plant.

4. Such revision will not contribute to the POTW's inability to comply with its NPDES permit, sludge use or disposal regulations (section 405 of the Act), or other applicable regulations, criteria, etc.
5. Approval of removal allowances requested will not result in the POTW violating water quality standards.
6. Grants from funds authorized from September 30, 1978 must have completed the analysis required by the act and demonstrate that the removal allowance claimed will not preclude the use of alternative or innovative technology or interfere with sludge or effluent reuse.

Supportive Data

Supportive data and information supplied to the approval authority in consideration of the POTW's application to revise discharge limits must meet the following requirements:

1. The pollutant(s) for which modification is requested.
2. Operation data must be presented on influent and effluent and any other information which demonstrates consistent removal for specific pollutant(s) in question. These data presented must:
   a. Represent yearly and seasonal conditions;
   b. Represent quality and quantity of normal effluent and influent flows; if such data is unobtainable, alternative data or information may be presented; wet weather projections, if applicable, should be included;
   c. Be obtained through composite sampling according to section 403.7 (c)(2) of 40 CFR. The approval authority may require an alternative sampling schedule. Grab samples must be used where composite sampling is not appropriate (e.g., cyanide, phenol, etc.);
   d. Be based provisionally on treatability studies or demonstrated removal at other treatment facilities where influent is similar for those pollutants which are not currently being discharged and where application for provisional authorization to revise discharge limits is desired.
3. Analysis performed in obtaining data presented for modifying standards should be conducted according to procedures prescribed in the particular standard, 40 CFR, Part 136 and amendments, and/or the EPA publication, *Sampling and Analysis Procedures for Screening of Industrial Effluents for Priority Pollutants, April 1977.* In some cases, it is anticipated that standards may not require direct analysis of priority pollutants, but will utilize indicators such as chemical oxygen demand (COD), total organic carbon (TOC), etc.
4. A list must be provided of industrial categories and number of users in each subcategory, and pollutants must be identified.
5. Alternative pretreatment limits must be proposed.
6. Data must be provided showing concentrations in POTW sludge (using sampling techniques similar to those used for liquid stream).

7. The POTW's current sludge disposal methods must be specifically described, and data indicating continuing compliance must be included.
8. It must be certified that the pollutant removal efficiencies and alternative pretreatment limits are calculated in accordance with EPA regulations and guidelines.

The following formula is recommended by EPA in calculating the proposed pretreatment limit for a specific pollutant:

$$Y = \frac{X}{1-r}$$

where

Y = revised discharge limit for specified pollutant (mg/l)
X = pollutant discharge limit specified in the applicable categorical pretreatment standard (mg/l)
r = consistent POTW removal rate percentage expressed in decimal form.

Variances From Categorical Standards: Fundamentally Different Factors

In establishing categorical pretreatment standards for existing sources, EPA will take into account all the information it can collect, develop and solicit regarding the factors relevant to pretreatment standards under section 307 (b). In some cases, information which may affect the pretreatment standards will not be available or, for other reasons, will not be considered during their development. As a result, it may be necessary on a case-by-case basis to adjust the limits in categorical pretreatment standards, making them either more or less stringent, as they apply to a certain IU within an industrial category or subcategory. The modified standard will, in effect, acknowledge that the IU in question is fundamentally different from other facilities considered by EPA in developing the categorical standards.

Any interested party (including the IU) believing that the factors relating to an IU are fundamentally different from the factors considered during development of a categorical pretreatment standard applicable to that user and, further, that the existence of those factors justifies a different discharge limit from that specified in the applicable categorical pretreatment standard, may request a FDF variance or such a variance request may be initiated by EPA.

Where an NPDES state has an approved pretreatment program, it will be responsible for receiving requests for FDF variances and denying such variances where appropriate. Authority to approve such requests is reserved for EPA, but the state will have authority to recommend approval of the variance to EPA.

Upon receipt of a FDF variance request, the state should review the request for completeness. Where the state determines that the request is not complete or that its substance is in some way deficient, it should give the requester thirty days to make the appropriate corrections. The state may extend the time for making corrections at its discretion. If the requester does not make corrections within the time period specified by the state, the state should deny the requests and so notify the requester.

Requests for variances with supporting evidence must be submitted in writing to the state. Requests must be submitted within ninety days after EPA promulgates the particular categorical pretreatment standards. Requests submitted after the ninety-day period will not be considered. The written variance submission is required to contain the following information:

1. Requester's name and address;
2. Requester's interest in variance;
3. POTW receiving industrial waste;
4. Applicable categorical pretreatment standard;
5. List of pollutants or parameters in question;
6. Alternative limits proposed for each appropriate pollutant or parameter;
7. Description of the existing industrial water pollution control facility;
8. Schematic of industrial water system—water supply, process systems, and discharge point;
9. Clear statement of need for variance including supportive data; documentation, and appropriate evidence (such as EPA development documents, economic data, legal proceedings, etc.);
10. Where control technology recommended will not meet the standard, or exceed the standard, it is essential that adequate documentation be presented.

A request for variance must be reviewed in detail and it must be determined whether or not the factors identified are fundamentally different. The following points are essential for approval of a variance from a categorical pretreatment standard:

1. Factors relative to the affected IU are fundamentally different from the factors EPA considered in establishing the standard. Factors which may be considered as fundamentally different include:
   a. Fundamental aspects of the industrial process which significantly affect the nature or quantity of the process wastewater and discharge;
   b. Age, size, land, and space consideration as related to equipment and facilities, processes utilized, and application of control technology;

    c. Adverse impact on environmental quality other than water quality by control technology required to meet a categorical standard;
    d. Increased energy requirements, provided less energy-consumptive control technology is not available or applicable;
    e. Disproportionate compliance costs due to one or more of the above.
2. The factors upon which the variance is based existed prior to EPA promulgation of the particular standard.
3. Because of the fundamental difference in question, the cost of compliance with the categorical pretreatment standard would be grossly disproportionate to the cost of compliance EPA considered in establishing the standard.
4. The alternative pretreatment limits are justified by the extent of the fundamental difference.
5. The request for variance is in accordance with the procedures in the regulations.

The determination on the FDF variance should be made by a technical professional, such as a sanitary or environmental engineer. If, after reviewing the FDF variance request the state determines that it should be denied, the state should issue a written statement to this effect to the requester and to the IU in question if it is not the requester. This statement should outline the reasons for not granting the request.

If the state decides that the request should be approved, it should forward the request, along with a recommendation of approval and reasons supporting this recommendation to the EPA regional enforcement division director.

The pretreatment regulations allow for an appeal procedure for FDF determinations made by EPA. The regulations do not require that states develop such procedures for state determinations to deny a request. The denial of a request by a state cannot be appealed to EPA.

The following are estimates of the level of state effort required to apply the procedures previously outlined:

1. Initial cursory review of FDF request for completeness;
2. Insufficient request notification if required;
3. Determination as to whether or not an FDF exists.

SLUDGE CRITERIA

Improving the opportunities to recycle and reclaim sludges and effluents from wastewater treatment facilities is one of the intents of the national pretreatment policy. Removal credits should be approved where sludge management techniques are consistent in meeting this objective.

Consideration of future sludge management requirements via section 405 of CWA should play a major role in determining the acceptability of providing removal credit. The POTW should assure the approval authority that any additional amount of incompatible pollutants allowed into the POTW, due to approved removal credits being granted, will not contaminate the sludge or otherwise interfere with its use or disposal.

Under section 405 of the Act, the determination of the manner of sludge use or disposal is a local decision. However, once a sludge management option is selected, it is unlawful for the POTW to violate federal guidelines established under section 405 for that use or disposal option. The POTW's NPDES permit will include applicable guidelines from section 405 as they are available. It will be good business on the part of a POTW not to accept pollutants that will create a sludge management problem.

At this point in time, it appears the approach EPA will take in reference to CWA and RCRA in resolving the municipal sludge situation is as follows:

1. Municipal sludges will not be covered under RCRA regulations.
2. Section 405 regulations will be as (or more) restrictive as the requirements developed under section 4004, 3004, for RCRA. The regulations required by section 405 of CWA are expected to be out in 1980.

Disapproval of a removal credit submission, where the POTW sludge will be in violation of pending 405 regulations, will result unless compliance can be established in a reasonable amount of time (12 months).

Section 405 of the Clean Water Act and the Resource Conservation and Recovery Act and other applicable sludge disposal considerations are discussed in more detail in Chapter 5.

REFERENCES

1. General pretreatment regulations for existing and new sources of pollution. *Federal Register,* vol. 43, no. 123 (June 26, 1978).
2. *Local Pretreatment Program Requirements and Guidance.* Springfield, Virginia: Environmental Technology Consultants, Inc., June 1979.

3

Elements of a Pretreatment Program

BACKGROUND

The objective of a pretreatment program within a POTW system is to ensure continuity of treatment, provide physical protection of collection and treatment facilities, and prevent the discharge of pollutants from the treatment plant which would violate NPDES permit conditions or other regulatory requirements. Recently, additional impetus has been provided for the establishment of viable local programs by the general federal pretreatment regulations, which require local enforcement of categorical pretreatment standards. In order to achieve these objectives and comply with regulatory requirements, POTW operators must develop comprehensive programs which address the complex technical, economic, legal, institutional and environmental factors inherent in industrial waste control.

This chapter presents a detailed review of the five most essential elements of a pretreatment program, namely the development of a data base, the legal aspects, administration, monitoring and reporting, and user charge systems. Each element is discussed in detail, with specific examples and case histories included to illustrate the principles involved.

DEVELOPMENT OF A DATA BASE

The development of an accurate data base upon which a pretreatment program can be established is perhaps the most significant step in the evolution of a viable program. The information obtained in this exercise forms the basis for all of the critical steps which follow in the development and management of an overall program. An accurate data base is essential to

27

the preparation of specific stipulations in the ordinance for industrial use of sewers, and for the user charge system. It also provides the information which may frequently be necessary to establish the proper organization for administration of a pretreatment program. Finally, it is of significant importance in the eventual enforcement of the provisions of the industrial waste control ordinance.

The development of a data base is accomplished by means of an industrial waste survey, which can be approached in a number of distinct ways. Common to all approaches is the identification of the industries using the system, and the characteristics and quantity of their effluents. The general pretreatment regulations require that the POTW "identify and locate all possible Industrial Users which might be subject to the POTW Pretreatment Program" and make the inventory of industrial users available to the EPA regional administrator or state director. The regulations have similar provisions regarding the "character and volume of pollutants contributed to the POTW by the Industrial Users identified"(1).

The determination of wastewater flow and characteristics can be accomplished in one of three ways, ranging from the most simplistic approach to a complex and rigorous sampling program. The first and most simple method estimates the specific industrial contributions, based on waste characteristics published in the literature. The second method goes one step further, requiring that all significant contributing industries submit analyses of their waste streams to the POTW for review. The third method differs from the second in that the POTW itself samples and analyzes each discharge from a significant contributor.

The third method is always preferred, in that it will provide the most accurate and unbiased data. By scheduling sampling and analyzing the industrial waste stream itself, the POTW minimizes the possibility that the results will be obtained on atypical production days or that some of the discharge and flow data will be omitted. Although this method requires the POTW to commit the most resources to the survey, in the final analysis the investment will be more than justified by the accuracy and reliability of the data obtained.

In actuality, a combination of all three methods might be used to good advantage by a POTW embarking on a pretreatment program. Published data from the literature can be employed as an initial estimate, and subsequently used to compare the discharge from a specific plant with the wastewater characteristics of typical plants in the industry. Similarly, self-monitoring data can provide a second set of data which can then also be compared with the results obtained by the POTW in its own sampling program. In summary, all available applicable data from any reliable source should be utilized to establish the most complete, accurate, and reliable data base on the industries within the POTW system.

Regardless of which method or methods are ultimately employed, an industrial waste survey is invariably a phased operation. An example of an approach to the survey is shown in Figure 3-1 which depicts the development of the industrial waste control program for the Buffalo Sewer Authority (BSA) in Buffalo, New York (2). The industrial waste survey segment of the program is outlined by a dashed line in the diagram. In this instance, a three-phase approach was used, consisting of identification, cross section sampling, and follow-up sampling. In the second phase of the program, a cross section of the industries in the BSA system was sampled. The third phase constituted a more detailed sampling effort. Typically, an industrial waste survey consists of five phases. It must begin with identification of the industrial contributors; it then proceeds through a number of specific steps, namely, preliminary data analysis, a questionnaire survey, detailed data analysis, and a sampling program. The following paragraphs provide specific details on conducting an industrial waste survey, utilizing the typical five-phase approach.

Identification of Industrial Contributors

The objective of the initial task in the industrial waste survey is the compilation of a master list of industrial and commercial contributors within the POTW system. In developing the inventory, the approach should be to make the list as all-inclusive as possible, since insignificant dischargers can always be omitted at a subsequent point in time. Factors of interest for the initial identification include location of the industrial facility, product line, production volume, indication of whether the industry is wet or dry, and if wet, its water usage.

In the identification phase, a system of industry classification should also be established. It is advisable and practically essential that the standard industrial classification (SIC) system developed by the Federal Office of Management and Budget be used for this purpose. This system delineates industrial categories and subcategories by numerical designations. The SIC system has been adopted by the Environmental Protection Agency, and is used to identify the industries covered by the categorical pretreatment standards. Consequently, it is recommended that POTWs employ the same system for industry classification in order to be consistent with the EPA approach.

For the small municipality, identifying contributors within the wastewater collection system is usually an easy task. The plant operator, or the system's superintendent, is generally familiar with the area served, and the contributors within it. For larger systems, and those small systems where this familiarity does not exist, identifying industrial contributors can be a more complex task. The location of wastewater dischargers can be determined using various sources of information, including publicly available listings of

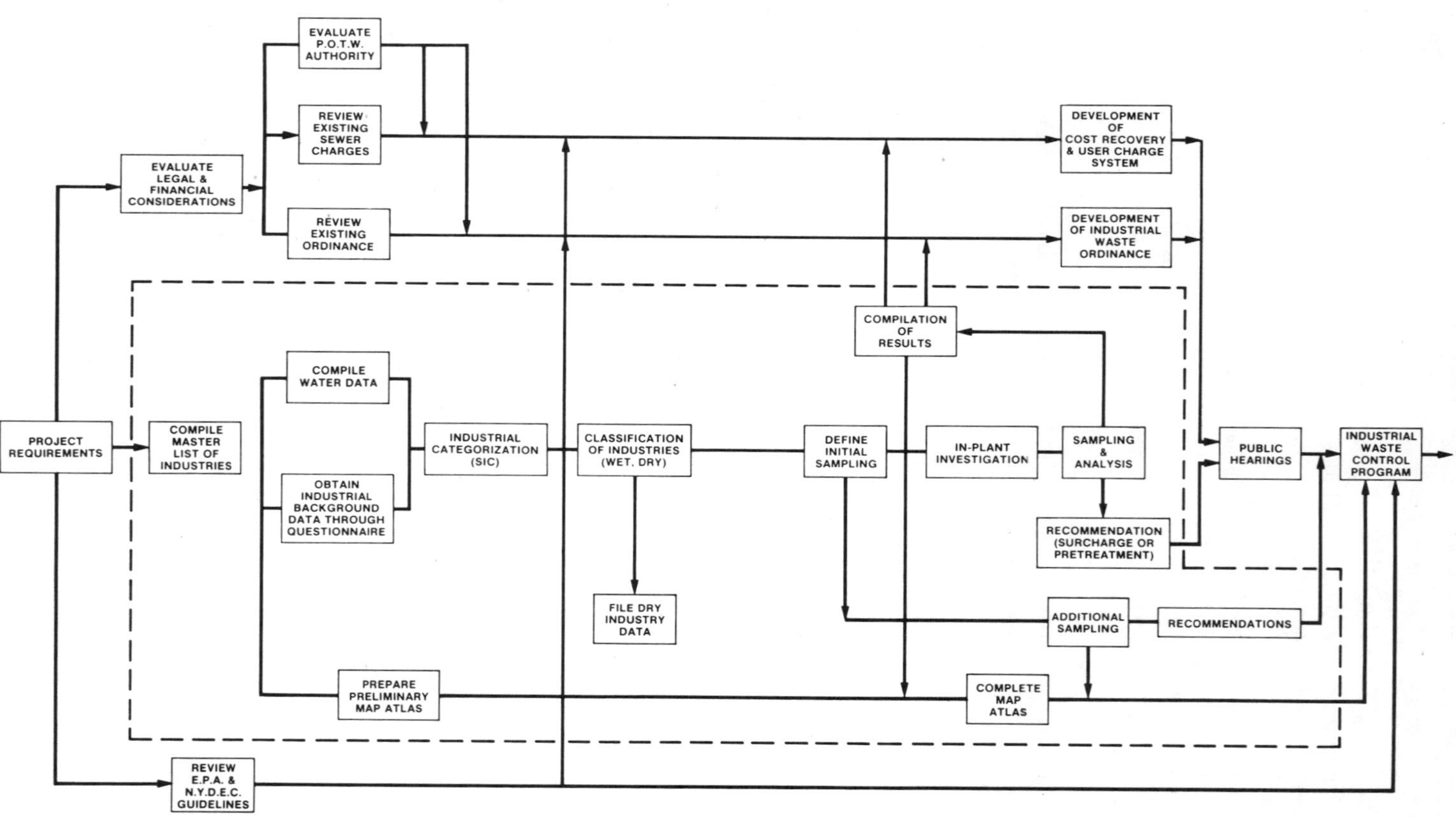

Figure 3-1 / Development of an industrial waste control program.

industrial and commercial establishments. The most effective sources of information include the following:

—Existing sewer authority files;
—City and state industrial directories;
—Labor department records;
—Property tax records;
—Chamber of commerce rosters;
—Census Bureau records;
—Local telephone directory;
—Water consumption records;
—Dun's Market Identifiers—Dun & Bradstreet.

The various directories, rosters and records have been compiled for different purposes, and consequently one information source may include an industry which another source has omitted. The review of all available information sources is recommended, so that the references can be utilized to cross-check each other and thus make the inventory as all-inclusive as possible.

Dun's Market Indentifiers (DMI) is a computerized source of information available from Dun & Bradstreet, Inc., which can selectively provide data on all industrial and commercial establishments in the United States. Available data includes the name of the company, its address, whether the location is a manufacturing or nonmanufacturing facility, the SIC code, sales volume, number of employees, name and title of the chief executive, and other related useful information. The DMI can provide an effective tool to be utilized as a final check on the other sources if information of the resources of the POTW permit its use.

Preliminary Data Analysis

The second phase of the survey consists of analyzing the data obtained in the identification phase, and utilizing it to serve as the basis for the questionnaire survey which follows. The master list of industries should be reviewed to eliminate dry industries from further consideration, and to tentatively identify the significant dischargers within the system.

A preliminary breakdown between major and minor contributors generally can be made on the basis of flow, which in turn can be estimated from water usage. In addition to consideration of flow, the municipality must decide, on the basis of specific other factors in the POTW system, which industrial facilities to consider significant contributors. A minimum flow value, such as 25,000 or 50,000 gpd, should be used as a first step in a major-minor delineation. However, other factors, including water quality standards and the POTW's sensitivity to a particular type of wastewater, must also be taken into account.

If the POTW is concerned with the federal provisions for industrial cost recovery and user charges, the selection of 25,000 gpd as a minimum flow for significant dischargers is suggested. This flow value is indicated in section 204 of the Clean Water Act of 1977 as a determining factor in the applicability of the industrial cost recovery and user charge programs. Consequently, for the sake of consistency, the same flow value should be utilized in the pretreatment program.

Establishing a reliable list of significant contributing industries on this basis ultimately requires direct contact with industry, and detailed analysis or evaluation of each plant's wastewater. Nevertheless, at this stage of the survey, any industrial discharge suspected of having such characteristics should be tentatively classified as significant.

The final list of industries to be included in the questionnaire phase of the survey should at the minimum include all significant dischargers; it will then automatically encompass facilities subject to federal pretreatment regulations. If POTW resources permit, the questionnaire should be mailed to the most inclusive list of industries possible, in order to preclude the possibility of missing any significant dischargers. A small additional investment for mailing costs and data processing at this stage of the program can help ensure the reliability and completeness of the data base.

Questionnaire Survey

This phase of the overall industrial waste survey is essential to confirm the data previously obtained from the various sources of information used in the identification of industrial contributors. The development of a questionnaire form should consider the degree of detail desired, which is usually related to the size and complexity of the POTW system. At the minimum, information should be requested on production, employment, water usage, the number and location of sewer connections, wastewater discharge characteristics, and historical analytical data. If section IV of NPDES standard form A has not yet been completed for each major discharger in the system, it can be conveniently attached to the questionnaire.

It is essential that sufficient attention be given to the details of the questionnaire to avoid any ambiguity and confusion, and to ensure that it includes questions relating to all the information that is desired. A recommended procedure is to try out the preliminary questionnaire form on a few selected industries to ensure that it is easily understood, and to incorporate any valid recommended changes or additions into the final form.

Figures 3–2 and 3–3 illustrate the first two pages of the questionnaire utilized by the Buffalo Sewer Authority (2). The third page of the BSA

Sic # _______________
Map# _______________
Loc# _______________
S.D. _______________
(For BSA use only)

Company Name____

Address____

Representative___________________________Title______________________Phone No.__ _______

Hours of operation/day___________________Days of operation/week___________________

No. of employees: Shift No. 1___________Shift No. 2__________Shift No. 3__________

Type of Business: (Manufacturer, Distributor or Retail) _______________________________

RAW MATERIALS	AMOUNT PER YEAR
_________________________	_________________________
_________________________	_________________________
_________________________	_________________________

PRODUCTS	AMOUNT PER YEAR
_________________________	_________________________
_________________________	_________________________
_________________________	_________________________

Type of Process: Continuous___________________   Batch___________________

Industrial Wastes:

What waste products are disposed to: Sewer___________________ Other___________

Is discharge to sewer: Intermittent__________   Steady__________________

                    Quantity/day__________   Est. or measured____________

Are wastes pretreated? If so, which and how:_______________________________________

Plant Sewer Connections to BSA

	Size & Shape	Material	Location in Plant	Connected To
(1)				
(2)				
(3)				

Are maps showing sewer connections available?__________

ADDITIONAL INFORMATION TO BE SUPPLIED ON YOUR LETTERHEAD.

Figure 3-2 / Buffalo Sewer Authority: Industrial waste survey (sheet 1 of 3).

ANNUAL VARIATION IN OPERATION

Is there a scheduled shutdown?_____________________   When?_____________________

Is production seasonal?_____________________

If so: Period of full production   _____________________to_____________________

                Period of limited production_____________________to_____________________

                Period of no production   _____________________to_____________________

                Employees (No.)   Max._____________   % of time at max._____________

                                  Min._____________   % of time at min._____________

If not: Average # of employees_____________

WATER USE

Source(s) of water____

If from an agency, Account #____

Water used for:

                                                                      Recirculated
                Sanitary   _____________ gpd

                Air Conditioning   _____________ gpd   _____________

                Process water   _____________ gpd   _____________

                Jacketed cooling water   _____________ gpd   _____________

                Other____

Period of max. water use_____________________   Amount_____________________

Period of min. water use_____________________   Amount_____________________

Water disposal other than sewer_____________________   % of total_____________________

Is water consumed in product?_____________________   Amount/day_____________________

Type and number of air pollution devices____

Have the waste streams been previously analyzed?_________________________________

Are radioactive isotopies used in your process?_________ Specify:_____________________

Figure 3-3 / Buffalo Sewer Authority: Industrial waste survey (sheet 2 of 3).

questionnaire, relates to the chemical analysis of the plant's discharge, and was only applicable if a previous analysis had been performed by the plant. This questionnaire form is included here to illustrate how a city the size of Buffalo accomplished this phase of the industrial waste survey. It may provide a starting point for another city facing this task, but it is important to note that the questionnaire must be tailored to the specific circumstances and needs of the POTW conducting the survey.

A public relations effort should be incorporated into any program if possible, making use of radio and television as well as newspapers to inform and solicit cooperation from the industries involved and the public at large. In addition to this effort, plans should be made for direct telephone contact with a major portion of the industrial community. Additionally, plant visits to industries which require help in completing the forms should also be planned as part of the program.

The Buffalo Sewer Authority's experience regarding responses to the questionnaire indicated that, at the very least, telephone followup is essential to obtaining meaningful information. They employed a newspaper campaign during this phase of the survey to acquaint industry with the program and to stimulate cooperation. BSA's experience was that approximately 40% of the establishments returned the questionnaire with some reminding by phone, 30% of the questionnaires were completed over the phone, and the remaining 30% of the questionnaires were completed by plant visits (2). In a similar survey conducted by the East Bay Municipal Utility District of Oakland, California, 60% of the industries receiving a questionnaire voluntarily returned the completed form to the District (3). Undoubtedly, this type of variation in response rate will be experienced, with the success of the questionnaire survey depending to some degree on the relationship between the POTW and the industries involved and the extent of the public relations program employed.

Detailed Data Analysis

Analysis of the data obtained through the questionnaire survey should be primarily aimed at delineation of those industries to be included in the sampling program. This task essentially involves confirmation and/or correction of the original industry data by means of a detailed analysis of the questionnaire responses. Preliminary information concerning wet vs. dry category, SIC classification, flow, production volume, etc. should be updated for each industrial facility submitting a completed questionnaire.

Another important task in this phase of the work is the review of wastewater data submitted by the industry. This information may in many instances constitute the first data available on the characteristics of the discharge from a specific facility. In municipalities where there has not been a previous enforcement program, industries in general know very little about their wastewater characteristics. The questionnaire program provides the incentive for them to discover the strength of the wastewater that they are discharging to the sewer system, and any problems and difficulties that are encountered in obtaining a representative sample of wastewater. These data also provide the basis for the POTW to decide which pollutant parameters require monitoring in the subsequent sampling program.

Sampling Program

The sampling program constitutes the last and perhaps most important phase of the development of a data base for industrial contributors to a POTW system. A subsequent section of this chapter is devoted to specific aspects of monitoring and reporting; however, several significant points regarding the sampling program must be emphasized here.

The first related to planning and scheduling, which should be thoroughly addressed prior to the start of the program in the field. It is essential that the initial sampling program obtain accurate, representative data, in that this information will form the basis for specific stipulations in the ordinance, development of the user charge system, and the establishment of a viable enforcement program.

Another important consideration in this phase of the effort is the role of self-monitoring in the development of a data base. Self-monitoring can be helpful in POTW systems where resources are limited. This type of approach is also most applicable where a local permit program is utilized, and the self-monitoring data form then becomes a permit application. Data accumulated in this manner can serve as the basis for establishment of a pretreatment program, with verification achieved through subsequent compliance monitoring. Nevertheless, it is recommended that the initial data-gathering effort be performed by POTW personnel whenever possible. Confidence in the data at this stage of development of the overall control program is essential, and this can best be achieved by means of a POTW sampling effort. The investment in personnel and other resources to accomplish this task can ultimately be recovered from industry by various cost recovery techniques available to most POTWs.

As indicated, the sampling effort should preferentially focus upon the pollutants identified by each industrial facility in the questionnaire survey. However, the initial sampling effort by the POTW should serve to screen each discharge that is monitored to confirm the presence or absence of a specific pollutant, as well as to obtain further data on the pollutants known to be present. This is especially important as a result of the emphasis being placed on the priority pollutants delineated in the Clean Water Act of 1977. If possible, the POTW should allocate the resources necessary to accomplish this screening sampling effort for at least all of the significant dischargers in the system.

LEGAL ASPECTS

The importance of the legal aspects of a pretreatment program cannot be overstated. Legal considerations play a major role in the development of an effective program and its implementation on a day-to-day basis. Specifically, legal aspects encompass three major areas of concern; the legal authority of the POTW over industrial dischargers to its sewer system; the ordinance to control such discharges; and enforcement of the ordinance. The establishment of legal authority and preparation of a workable ordinance are integral parts of the development of a pretreatment program, whereas enforcement is an essential element of the program once it has been implemented. Consequently, legal assistance must be viewed as a continuing need which does not end once the ordinance is drawn and instituted. Proper enforcement requires teamwork among administrators, engineers, attorneys, and field and laboratory personnel, with the legal role being the key to eventual resolution of problems through conciliation or court action.

Legal Authority

One of the most significant factors in the establishment of a viable pretreatment program is the provision of adequate legal authority to develop, administer, and enforce the program. Whatever agency is designated to operate the program must have sufficient power to enforce its rules and regulations on industrial users, and to obtain the data necessary to monitor how its rules are being complied with.

In many cases, legal authority may be complicated by the structure of a sewer district or regional sewer authority. Many districts and regional authorities act essentially as wholesalers serving a number of political subdivisions, and do not have any direct contact with industries in the system. In these POTWs, the industries discharge to municipal sewers which in turn connect to the regional agency. In such instances, the development and enforcement of ordinances controlling the discharge of industrial wastewaters is legally the responsibility of the individual municipalities in the system. They may look toward the regional agency for guidance in such matters, but the municipalities retain the legal authority to deal with industries in their systems in these situations.

The problems inherent in the type of situation described are numerous. Frequently, small municipalities whose sewer systems are connected to regional POTWs do not have the resources or inclination to control the industries within their system. The POTW may suffer an upset in treatment plant operations, and even a resulting violation of its NPDES permit, as a result of an industrial discharge originating from one of its municipal collection systems. In such instances, the POTW must have the legal authority to directly and immediately control the industrial discharge to protect its operations.

Recognizing the need to correct situations of this type, the revised general pretreatment regulations (1) require the POTW to obtain legal authority to apply and enforce the requirements of sections 307(b) and (c) and section 402(b)(8) of the Clean Water Act. As a minimum, this legal authority must enable the POTW to:

—Reject or impose conditions on new or increased discharges of pollutants to the POTW by industrial users;
—Require compliance with pretreatment standards by industrial users discharging wastes to the POTW;
—Control, through permit, contract, order, etc., the discharge to the POTW by each industrial user to ensure compliance with pretreatment standards;
—Require the development of a compliance schedule by the industrial user to meet applicable pretreatment standards and the submission of notices and self-monitoring reports from industrial users;
—Carry out inspection, surveillance and monitoring procedures necessary to determine compliance with applicable pretreatment standards, including authorization to enter the premises of any industrial user in which an effluent source, records, or a treatment system is located;
—Independently assess, or recover through judicial action, civil or criminal fines, penalties, and injunctive relief for the noncompliance by an industrial user. The POTW shall also have authority to halt or eliminate any discharge of pollutants to the POTW which presents an imminent or substantial danger to people or to the proper operation of the POTW.

The regulations state that the legal authority "may be contained in a statute, ordinance, or series of contracts or joint powers agreements which the POTW is authorized to enact, enter into or implement, and which are authorized by state law." POTWs faced with the problem of obtaining legal authority, should have their legal counsel examine all aspects of their charter or compact to determine the most direct method for acquiring such authority.

The solution to the problem depends to some extent on the legal structure of the sewerage agency. If it is only a voluntary association of independent municipalities, the agency will generally only have such authority as has been delegated to it by the compact creating the sewage authority or district. If the sewage agency has been established by a superior governmental jurisdiction, it then may be able to superimpose its authority on that of the local municipalities.

In some instances, it may be necessary to have the state legislature revise the compact or charter of the regional agency in order to obtain legal authority. In other instances, renegotiation of the agreement between the regional agency and the municipalities in the service area might be the most direct method for obtaining the necessary authority. In some particular situations the 208 planning agency may be able to provide assistance in obtaining the legal authority that the POTW needs to interface directly with industrial contributors to the system. In any case, to comply with NPDES permit and pretreatment requirements, it is essential that the operating agency of the POTW obtain the power to deal directly with industry in order to establish a viable pretreatment program.

Enforcement

POTW control agencies will play the major role in ensuring and enforcing compliance with pretreatment standards on the part of industrial users. Where approved POTW pretreatment programs exist, state and federal enforcement activities will serve in a backup capacity. The POTW control agency derives its enforcement powers from the legal authority established in developing the program.

The approval authority may request reviews of the industrial users' self-monitoring reports in order to spot check the credibility of the POTW control agency in detecting violations. The approval authority will exercise its enforcement authority when necessary, to correct inefficiencies or insufficient enforcement at the POTW level. Actions will be taken against a POTW control agency for inaction against an industrial discharger in violation.

The CWA, besides granting enforcement mechanisms and authority to EPA and NPDES states, also provides, through section 505, the authority for citizens to bring civil action against violation of pretreatment standards. It is essential to eliminating future problems that this consideration be taken into account when developing a POTW pretreatment program and subsequent public notification.

The main vehicle for enforcing POTW pretreatment programs will be through the POTW's NPDES permit. Compliance schedules for developing a pretreatment program will be incorporated into the permit. This schedule will not exceed three years. Other conditions enforceable through the permit include:

1. Executing adequate monitoring, reporting, and inspection activities;
2. Enforcing the standards when violation occurs;
3. Maintaining a demonstrated percentage removal of pollutants where removal credits have been granted;
4. Ensuring that sludge management practices do not violate applicable criteria; and
5. Maintaining proper reporting for permit compliance.

In addition, EPA has the authority to enforce against a POTW for pretreatment violations by its industrial users even if there is no POTW pretreatment program enforceable through the POTW's permit.

It will prove to be a good business technique for municipalities to:

1. Develop sound pretreatment programs;
2. Enforce the requirements; and
3. Continue POTW operation to maintain removal credits and applicable sludge management.

Sewer Use Ordinances

The ordinance is the heart of any pretreatment program, providing the essential mechanism for controlling the discharge from industries within the POTW system. Consequently, the ordinance must be carefully drawn to include all essential ingredients for the particular system. Once promulgated, the ordinance should be utilized for the control of all industrial wastewater discharges and the eventual enforcement of its terms against all violators. Therefore, the administration or enforcement of the ordinance is just as important as its content.

Historical Perspective

The content of sewer use ordinances has evolved over the years in a way roughly paralleling the increase in sophistication of wastewater treatment facilities. Early ordinances were outgrowths of local plumbing and health codes, the contents of which were primarily devoted to standardization of materials used in the construction of sewers and connections thereto. Since sewage treatment was limited to settling of solid material, the content of ordinances at this stage focused on protection of sewers from clogging, corrosion, and explosive hazards. Most ordinances are built around these basic provisions and have increased in scope as new needs have arisen.

With the advent of secondary treatment processes and the development of anaerobic digestion, closer control over certain dissolved organic and inorganic pollutants became necessary to prevent inhibitory effects in these units. In many instances, this type of control was provided by setting concentration limits for the acceptable discharge of critical pollutants.

The recent application of state and federal effluent and water quality standards, which in many cases requires the removal of even trace quantities of certain toxic pollutants, has caused many POTWs to establish direct control over significant industrial users. To gain information and control over these sources, many ordinances are now including permit provisions and self-monitoring provisions, or both (4).

Types of Ordinances

There is no single ordinance now in force that could be considered typical of all, or even most, ordinances currently being utilized. Effective control of sewer use has been obtained by both simple and complex ordinance structures.

Two factors influence the size and content of a sewer use ordinance. The first of these is whether the ordinance is designed to be self-contained, or whether it simply states general provisions and relies on separately published rules and regulations for interpretation and implementation. The second factor is whether or not the ordinance incorporates a permit system for either industrial users or all users of the system.

Obviously, an ordinance which is self-contained will be a longer document than one which is not. Permit systems, if used, will also add length. Generally, smaller communities with relatively few industries will use a self-contained ordinance and will not employ a permit system. Enforcement of the ordinance for the few industrial users in such cases can be achieved on the basis of personal contact between the pollution control officer and the industry plant manager. As the size of the POTW increases or, more importantly, as the number of industrial users increases, it becomes more difficult to provide flexibility for the numerous individual differences between users within the ordinance document itself. In these systems, a shorter document stating general provisions may be more effective. This type of ordinance is usually supplemented by a separate set of rules and regulations that explain the responsibilities of users with respect to the general provisions (4).

The short form of ordinance, with supplementary rules and regulations, is used by the Boston Metropolitan District Commission, the Metropolitan Sanitary District of Chicago, the Milwaukee Sewerage Commission, as well as the Metropolitan Cincinnati Sewage Disposal Program. The chief engineer is designated in all four cases as the responsible administrative official. The completely self-contained type of ordinance is used successfully by the Los Angeles County Sanitation Districts (5).

Regardless of the type of ordinance employed, its scope and complexity should be limited to the resources available to the enforcement authority. If enforcement of the ordinance relies heavily on a permit system and self-monitoring, then there should be an adequate office and field staff to issue and review permits and to check the periodic self-monitoring information. If enforcement is based on analysis of samples taken by the authority, then an adequate field and laboratory capability is necessary. In all cases, sufficient manpower must be available to comply with the requirements of the pretreatment regulations and to follow up violations through the administrative procedures provided in the ordinance. Only by following up on each violation to gain compliance will credibility for both the ordinance and the enforcement authority be established.

Ordinance Provisions

Two general sewer use ordinances are included in the Appendix to illustrate the structure, language and essential elements of this type of ordinance. The first is the model ordinance published by the Water Pollution Control Federation (6); while the second is the "Recommended Ordinance for Industrial Use of Publicly Owned Sewerage Facilities," excerpted from the federal pretreatment guidelines (4). The primary difference between the two ordinances is that the WPCF model ordinance is more inclusive, serving as a general sewer use ordinance. The pretreatment guidelines ordinance assumes that the general provisions dealing with requirements to connect, private sewage disposal, standards for construction of sewers and appurtenances, provisions relating to the control of infiltration and inflow and other miscellaneous provisions are covered elsewhere. Thus, it pertains only to those portions of a complete sewer use ordinance which relate to the industrial use of sewer systems.

Both ordinances are designed to be self-contained, and in their full form could be used independently as separate documents. In addition, individual sections of each ordinance may be used separately as an addendum or a major revision to an existing ordinance. Finally, individual clauses contained in both ordinances may be used to supplement an otherwise complete document.

In any case, the ordinances are included herein for reference purposes only, to illustrate application of the principles described in this chapter. Neither ordinance is intended to be utilized for any specific purpose merely by filling in the inserts. Both ordinances would have to be tailored to meet the needs of the specific local situation. A POTW faced with the need to promulgate an ordinance to control industrial discharges into its system would be well advised to carefully review both ordinances as examples of two approaches to the problem. The discussion which follows relates to the contents of both sample ordinances, but in a general sense describes the necessary components of any ordinance intended to regulate industrial wastewater discharges to municipal sewer systems.

The WPCF model ordinance contains seven articles, while the pretreatment guidelines ordinance has ten major sections. The various elements of both ordinances can be grouped into four essential parts as follows:

1. Introductory sections, consisting of the introduction, purpose, legislative background, statement of the area of jurisdiction, and definitions;

2. The body of the regulation, consisting of prohibitions and limitations on discharges, control of prohibited wastes, sampling and analysis of wastes, and enforcement procedures;
3. Optional add-on sections for a local permit program, user charge system, etc.;
4. Procedural clauses.

The introductory section generally contains a simple, short statement outlining the content of the ordinance. It usually includes the purpose, legislative authority, and precise location (city, county, state) of the area under jurisdiction of the ordinance. The introductory portion of the two sample ordinances represent differing approaches, with the WPCF model ordinance containing a terse statement of the content of the ordinance, while the pretreatment guidelines ordinance includes a statement of purpose and legislative background in addition to the content.

All ordinances should contain a section on definition of terms used in the body of the ordinance. As few as 10 and as many as 50 definitions have been used, depending on the sophistication of the ordinance. The sample ordinances contain virtually the same number of definitions, 22 and 23. The important aspect of the definitions section is that it should include all terms used subsequently in the ordinance that are not self-evident, and that may cause any misinterpretation or misunderstanding.

The body of the ordinance should contain all of its essential provisions, including regulations controlling discharges to the sewage system, requirements for sampling and analysis, authority and procedures for control such as submission of plans, admission of POTW personnel to industrial property, etc., as well as procedures for enforcement of the ordinance. The WPCF ordinance basically covers all of these elements in article V (use of the public sewers), with article VII providing for the powers and authority of inspectors. The pretreatment guidelines ordinance is structured so that section 2 (prohibitions and limitations on wastewater discharges), section 3 (control of prohibited wastes), section 4 (industrial wastewater monitoring and reporting), and section 6 (enforcement procedures), constitute the body of the ordinance.

In any case, the section or article covering prohibitions and limitations on wastewater discharges is the heart of the ordinance. Most ordinances cover prohibitions and limitations of wastewater discharges in two parts; general prohibitions on materials which have proven to be hazardous or interfere with both collection and treatment systems, and limitations on certain critical pollutants which either interfere with, or pass through the treatment facilities. The general prohibitions should delineate all objectionable materials as specifically as possible, and also should provide legal coverage for unanticipated problems. In this context, it should provide the

POTW with the flexibility to effectively act against violators discharging materials not specifically named in the ordinance.

One instance in which flexibility is essential is in regard to excessive discharge rate, or so-called slug discharges. Both sample ordinances define such discharges in the same manner, and prohibit wastewaters containing high concentrations of pollutants or unusually high flows which would cause a treatment process upset and subsequent loss of treatment efficiency. This clause is of particular importance to many POTW systems. In a number of instances, industrial dischargers constitute a large percentage of the flow to a POTW. Slug discharges can be extremely damaging in such cases. Excessive discharge can also be a significant problem with industries having wide seasonal variations in their plant effluent. In any case, it may be possible for a POTW to define a slug discharge more specifically for the system in question. This may be done in terms of design capacity, NPDES permit conditions, or by taking into account the most cost-effective means of dealing with excessive discharge. If it is possible to delineate specific limits for slug discharges incorporating local conditions, that is generally more desirable than the general type of clause contained in the sample ordinances.

The establishment of specific limitations for pollutant parameters is one of the most critical portions of any municipal ordinance for the control of industrial wastes. The limitations for specific pollutants may be handled in a general way in the ordinance, as in option A of section 2(b) of the pretreatment guidelines ordinance; they may be specified elsewhere, such as separate rules and regulations, as implied in article V, section 4(e) of the WPCF model ordinance; or the limitations may be incorporated directly in the ordinance, as in option B of the pretreatment guidelines version.

The general limitations provide flexibility for systems which are unable to determine what specific limitations to establish. However, if sufficient information is available to select pollutant parameters and set specific numerical limitations, that option is invariably preferable. Setting such limitations requires thorough knowledge of the sewage system, data on industrial contributors, and familiarity with applicable water quality standards, federal pretreatment standards, and the provisions of the NPDES permit for the POTW. In establishing specific limits, there are four major factors to consider; (1) federal pretreatment standards, (2) passthrough of pollutants which would affect the POTW's NPDES permit or water quality standards, (3) inhibition of treatment processes, and (4) interference with sludge handling or disposal operations.

Of these factors, setting limits for industrial contributors on the basis of passthrough of pollutants is the most complex. The method required to establish limits for industrial contributors based on permit or water quality provisions is back-calculating, using the allowable discharge level in the standard as a starting point. This procedure requires knowledge of the removal capability of the treatment plant and the average plant influent concentration for the pollutant parameters of interest. In addition, information is necessary on dilution, background levels of the pollutant in question, and peak-to-average-flow ratios within the system.

Frequently, sampling of the plant influent and effluent is required to establish its removal capability for a specific pollutant. Additional sampling and flow measurement within the collection system may also be necessary to obtain all of the required data. One major POTW system, the County Sanitation Districts of Los Angeles County, has developed a formula for use in establishing limits on its industrial contributors based on allowable pollutant concentrations set by the state of California in the treatment plant discharge to the Pacific Ocean.

The Los Angeles County formula is as follows (4, 5):

$$L = PR\,[L_p - L_a\,(1 - S)]$$

where

$$L_p = L_e/(1 - E).$$

The terms used in these equations are defined below:

E = treatment plant removal efficiency.

L = calculated maximum concentration allowable in an industrial discharge (mg/l).

L_a = existing average yearly influent concentration to the treatment plant before source control (mg/l).

L_e = allowable effluent concentration at the treatment plant (mg/l).

L_p = calculated permissible influent concentration at the treatment plant necessary to meet the discharge limit for a specified POTW removal efficiency (mg/l).

P = assumed ratio of maximum to average concentration in industrial wastewater containing a given pollutant.

R = ratio of total wastewater discharge to the quantity discharged containing a given pollutant.

S = fraction discharged by controllable industrial sources.

In this formula, L is determined for each pollutant parameter for which an L_e has been established. The dilution ratio R is estimated by determining the total flow of all industries to be regulated for the pollutant in question, and dividing it into the total flow to the treatment plant. Values for E, the treatment plant removal efficiency for the pollutant parameter, should be based on operating experience of the specific treatment plant. The peak-to-average ratio P is an estimate of the ratio of the maximum concentration of a constituent in an industrial effluent to its average concentration for all industries discharging the pollutant. The more tightly a pollutant is controlled, the lower this ratio will be. The factor S is included to account for background levels in the system. If there is no background level for a particular pollutant, with industry providing the total concentration, then the value of S is 1.0. L_a is obtained from monitoring the pollutant parameters of interest in the treatment plant influent, and is included because the factor $L_a(1 - S)$ represents the background concentration of a specific pollutant which will not be controlled by the established limit.

The originators of the formula, the County Sanitation Districts of Los Angeles County, have utilized it to aid in determining allowable pollutant concentrations for industrial dischargers within the system. Table 3–1 provides a numerical illustration of how the formula was applied to the ocean standards established by the State of California for arsenic, cadmium, chromium, copper, lead, mercury, nickel, silver, zinc, and cyanide. The table is included here only as a guide for utilization of the formula rather than for the presentation of numerical information. The values shown for E and L_a represent the results of sampling at the treatment plant, while the R and P factors were determined from accurate records of industrial discharges to the system. The determination of S was based upon sampling of domestic wastewater in the Los Angeles County collection system, with the results as shown in Table 3–2. Industrial companies which were considered to be noncontributors for a specific pollutant were assumed to discharge that pollutant at the domestic level, unless other data were available. The estimated controllable fraction from industrial sources, S, was then found by subtracting the total domestic portion plus the noncontributing industrial portion from the known daily mass input of those pollutants to the treatment plant.

Any POTW which wishes to apply the formula to its own situation must determine applicable values for all of the factors in the formula based on the characteristics of the individual system. Although this approach requires extensive data gathering and knowledge of the POTW system, it does represent a systematic method for establishing specific numerical limitations for industrial pollutants based on the allowable concentration of each pollutant in the treatment plant effluent.

The foregoing discussion relates to the establishment of pretreatment standards based on the passthrough of pollutants. In a similar manner, numerical limitations may be established based on the inhibition of treatment processes or interference with sludge handling or disposal operations that may be caused by particular incompatible pollutants. A number of factors within a given collection and treatment system will affect the acceptable level of a specific pollutant in a specific treatment plant. Of the many factors to be considered, the type of treatment process utilized is of prime importance. In addition to the inherent differences between unit processes such as activated sludge and trickling filtration, other treatment considera-

Table 3–1 / Calculation of maximum allowable industrial discharge concentrations, County Sanitation Districts of Los Angeles County (4).

Constituent	L_e, Ocean plan effluent limit mg/l	E Removal efficiency at plant w/biological secondary treatment %	L_p, Permissible influent limit to plant $\frac{L_e}{(1-E)}$, mg/l	L_a, average treatment plant influent concentration, mg/l	R, Dilution ratio[1]	P, Assumed ratio of maximum to average concentration	S, Assumed fraction from controllable sources	L, Calculated maximum concentration allowable in industrial discharge: $L = PR\,(L_p - L_a\,(1-S))$
Arsenic	0.01	48	0.02	0.025	35	10	0.47	2.4
Cadmium	0.02	73	0.09	0.032	35	10	0.86	29.4
Chromium	0.05	77	0.22	0.887	14	5	0.88	7.9
Copper	0.2	76	0.4[2]	0.635	23	5	0.76	28.5
Lead	0.05	80	0.25	0.370	19	10	0.75	29.9
Mercury	0.001	84	0.008	0.0012	350	15	0.60	39.5
Nickel	0.1	53	0.23	0.295	23	5	0.85	21.4
Silver	0.01	69	0.032	0.010	70	10	0.73	20.5
Zinc	0.3	77	1.58	1.99	18	5	0.84	113.5
Cyanide	0.1	55	0.24	0.430	18	10	0.93	37.8

[1]Dilution ratios are based on the industrial waste inventories of the sanitation districts. They are obtained by dividing 350 mgd by the total industrial wastewater flow of all industries identified in the inventory as significant contributors of the particular contituent.

[2]The limit of 0.4 mg/l copper is based upon the toxicity relationship of copper to biological wastewater treatment processes and is more restrictive than the influent limit required by the Ocean plan effluent limit. The restrictive L_p value of 0.4 mg/l was used in lieu of $L_e/(1-E)$ in the calculation of L.

**Table 3-2 / Concentration of
pollutants in domestic wastewater (5)**

Constituent	Concentration, mg/l
Arsenic	0.014
Cadmium	0.005
Chromium	0.020
Copper	0.119
Cyanide	0.029
Lead	0.051
Mercury	0.0005
Nickel	0.031
Zinc	0.490

tions such as chemical addition and wastewater flow pattern strongly affect inhibition characteristics.

Additional factors to be considered in setting limits based on inhibition or interference are the natural dilution available from other contributors to the system, and background levels of the pollutant present in the collection system. Dilution is primarily a function of the size of the system, ranging from insignificant dilution in many small systems, to extremely high levels of dilution in very large systems. Several of the largest systems report that upset of treatment processes is extremely rare, regardless of the concentration of individual industrial contributions. The background level of a specific pollutant in a sewage system should also be considered in setting numerical standards based on inhibition. Normally this is a difficult determination, but information from systems without industrial contributors may be helpful. Also, monitoring of a section of the collection system without industrial contributors and monitoring of all industrial dischargers and plant influent levels, with subsequent calculations, will reveal background levels within the system.

In general, judgment is required in developing appropriate limitations based on inhibition of treatment processes, and all factors affecting the determination must be taken into account at the local level. A final consideration in establishing limitations based on inhibition is the incorporation of an appropriate safety factor into the numerical limits for each pollutant, to ensure that adequate protection is provided for the treatment process. In a fashion similar to that used for calculating the passthrough of incompatible pollutants, the maximum allowable concentration of inhibitory or interfering pollutants at the inlet to a POTW or a specific unit process can also be calculated.

Calculation of pollutant levels at the influent to a treatment plant that will inhibit a unit process within the plant can be determined using the following equation (4):

$$L_p = \frac{L_i}{(1 - E_p)},$$

where

E_p = removal efficiency of the unit processes upstream of the unit process of concern;

L_p = permissible influent limit to the treatment plant;

L_i = maximum allowable concentration in the influent to the unit process that will not inhibit that unit process.

Once L_p has been determined using this formula, concentration limits at the point of industrial contribution can be ascertained by substituting L_p for L_e in the Los Angeles County formula and calculating L. Both L_i and E_p are unique for each treatment plant and should be determined in advance before attempting any calculations.

For illustrative purposes, values of 1.0 and 0.1 mg/l for L_i have been assumed as the inhibitory levels of copper for the activated sludge and nitrification processes, respectively. Similarly, average removals of copper in primary and activated sludge plants have been assumed as 25% and 50%, respectively. These values are utilized in the following sample calculations.

For the activated sludge process, using 1 mg/l for L_i and 0.25 mg/l for E_p, since some copper will be removed in the primary clarifier prior to activated sludge treatment, the permissible copper concentration in the influent to the treatment plant is calculated as follows(4):

$$L_p = \frac{L_i}{(1 - E_p)} = \frac{1 \text{ mg}/1}{(1 - .25)} = 1.33 \text{ mg}/1.$$

For the nitrification process, 0.5 is appropriate for E_p, since a standard activated sludge plant will precede most nitrification processes. Using 0.1 mg/l for L_i, the permissible copper concentration to avoid inhibition of nitrification processes is then calculated as follows (4):

$$L_p = \frac{L_i}{1 - E_p} = \frac{0.1 \text{ mg}/1}{(1 - 0.5)} = 0.20 \text{ mg}/1.$$

This type of calculation cannot be directly applied to anaerobic digestion processes. In a digestor, inhibitory materials may be present as either aqueous species or as insoluble precipitates. High concentrations of potentially inhibitory materials, inorganics in particular, are frequently observed in digestor sludge with no apparent decrease in operating efficiency. This is explained by the fact that the high levels typically result from a concentration of insoluble inorganic materials, which do not impact on biological activity. A major concern in anaerobic digestor operation is the concentration of inhibitory materials in the supernatant transferred with the sludge solids. The complicating factor in attempting to complete calculations such as those shown above is that the chemical environment in a digestor, such as pH changes or the presence of sulfide, often causes the pollutant to precipitate before it can inhibit the process. Therefore, calculations of limits based on inhibition of anaerobic digestion processes require a detailed evaluation of each individual plant's operations, and cannot be adequately illustrated in general terms.

It is important to observe that the inhibitory concentration observed for activated sludge processes may be higher than that for nitrification or anaerobic digestion processes. Since the activated sludge process utilizes many different bacteria, it can more readily withstand the effects of an inhibitory substance by adapting to the changing environment. On the other hand, nitrification and anaerobic digestion processes rely on the performance of a single strain of bacteria, thus making these processes less adaptable to the presence of inhibitory pollutants.

In addition to the consideration of unit process inhibition, specific pollutant limitations may have to be based on the prevention of interference with sludge handling or disposal operations. Essentially, any inorganic material removed in a typical biological treatment process is contained in the sludge wasted from the system. Using the example of a 50% copper removal cited above for an activated sludge plant, it is possible to determine the level of copper expected in the raw sludge. Assuming 1.0 mg/l of copper is present in the influent to a 1-mgd plant, then the quantity of copper found in the sludge can be calculated as follows(4):

$$\text{copper in sludge} = (1\ \text{mgd})(1\,\text{mg}/1)(.50)(8.34\ \text{lb}/10^6\text{gal}) = 4.17\ \text{lb}/\text{day}.$$

Since a typical activated sludge plant produces about 20,000 gal daily of a sludge containing 1% solids per million gallons, the concentration of copper in the waste sludge may be calculated as follows(4):

$$\text{copper in waste sludge} = \frac{4.17\ \text{lb}/\text{day}}{(20,000\ \text{gpd})\ (8.345\ \text{lb}/10^6\ \text{gal})} = 24.9\ \text{mg}/1$$

The dewatering of this sludge to a typical 20% concentration prior to disposal would increase the copper concentration to 498 mg/l. This calculation demonstrates that for a relatively small concentration of copper in the plant influent, the final concentration in the sludge is relatively high. As a result, it may be necessary to establish limits on the level of pollutants entering the plant on the basis of the ultimate disposal of sludge. The example cited above shows that one of the most critical impacts of industrial contributions of incompatible pollutants may be interference with sludge disposal operations.

As with setting limits to avoid inhibition of anaerobic digestion processes, calculating limits to protect against interference with sludge handling and disposal methods is difficult. Factors such as digestor detention time, methods of decanting and mixing, and various other operational considerations may impact on the quality of dewatered sludge obtained. Consequently, factors of this type should be defined prior to attempting to calculate limits based on interference with sludge handling or disposal operations.

The foregoing discussion has outlined methods and considerations relative to establishing numerical limitations for specific pollutant parameters. As indicated, such limitations constitute the essential ingredient of any ordinance to control industrial discharges to publicly owned sewage facilities. Limitations set on the basis of passthrough, inhibition, or interference to satisfy local requirements, must be integrated with the federal categorical standards for specific industries. Additionally, the necessary administrative and enforcement procedures must be incorporated into the body of the ordinance to permit the POTW to establish and run the pretreatment program effectively.

Administrative clauses in the body of the ordinance should cover subjects such as regulatory actions, submission of plans for pretreatment facilities, proper operation and maintenance of pretreatment facilities, admission of POTW personnel to industrial plants and reporting of accidental discharges. In addition, several clauses in the body of the ordinance should be devoted to monitoring and record keeping, encompassing the type and frequency of samples required.

Both sample ordinances cover the bulk of these subjects in varying degrees and with varying specificity. The pretreatment guidelines ordinance generally provides more detail on these matters in section 3 (control of prohibited wastes) and section 4 (industrial wastewater monitoring and reporting). The WPCF model ordinance deals with these subjects in articles V and VII essentially in a more general manner, stating that individual rules and regulations established by the superintendent will delineate specific requirements.

The other major subject that should be addressed in the body of the ordinance is enforcement procedures. Generally, state or local regulating agencies should recognize that concentration-based limitations have some inherent deficiencies. Mainly, in cases where water and sewer costs are less than the cost of installing and operating pretreatment facilities, industrial discharges may be tempted to use dilution to comply with concentration limitations. Dilution could result from the addition of city water, noncontact cooling water, or relatively clean process water or storm water.

To detect diluters, the responsible authority should review industrial plans and specifications for pretreatment facilities, and use on-site inspection of pretreatment facilities and plant piping to determine that the appropriate pretreatment facilities have been installed, are being operated, and that there are no permanent connections for dilution water. Additionally, close surveillance of water meter readings and records can often pinpoint diluters. However, this type of surveillance is usually most effective when the concentration limits first become effective and are first enforced. With these safeguards, concentration limits will usually provide an effective and enforceable means of preventing interference with treatment processes or the pass-through of pollutants.

The section of the ordinance dealing with enforcement should provide a structured approach for the handling of violations, as illustrated by section 6 of the pretreatment guidelines ordinance. Generally, enforcement should be a step-by-step process, starting with notification in writing of the violation, which would generally be followed by conference and conciliation. If these steps are unsuccessful in solving the problem, then more formal proceedings should be instituted.

Formal action could begin with a show cause hearing, which provides the violator with an opportunity to show cause why an order should not be issued directing the discontinuance of the discharge. Show cause hearings are usually open to the public, and held before a hearing board composed of appropriate officials of the municipality, authority, or district. Upon submission of all evidence, the hearing board has the option of issuing an order for the cessation of the discharge within a specified time period. If an order is issued and not complied with, then court action is the next and final step in enforcement of the ordinance. This step-by-step approach to enforcement provides the maximum opportunity for the resolution of ordinance violations, without resorting to court proceedings until absolutely necessary.

The optional and procedural provisions constitute the two remaining portions of an ordinance regulating the industrial use of POTW facilities. Whereas the sections of the ordinance dealing with definitions, prohibitions, limitations, control procedures, monitoring and enforcement are required in every ordinance, sections addressing local permit systems and user charges are considered optional since they are not applicable to every POTW.

Generally, local permit programs are most appropriate for large systems, and to date have been used successfully by several large regional agencies. Permit programs greatly increase the paperwork associated with pretreatment programs. Although the onus for filing permits is placed on the industries in the system, the POTW must nevertheless scrutinize the permits for completeness, truthfulness, accuracy, etc.

The use of permits can be advantageous in instituting a pretreatment program in a large system. At the outset, industries must provide all relevant information on their discharges, thus reducing to some degree the effort required on the part of the POTW staff for the initial data-gathering task. Local permit programs offer an option which is systematic and thorough, but one which requires a well-trained, capable staff to administer. Where local permit systems are utilized, the ordinance should contain specific provisions detailing the procedures involved. Similarly, where user charges are imposed or other cost recovery methods are employed, the ordinance should specifically delineate the formula or other basis for calculating costs.

The procedural clauses may serve a number of functions, such as referencing other related ordinances, reconciling differences that may exist between various ordinances, and establishing the effective date of the ordinance. In addition, many ordinances contain a "saving" clause, which provides that the bulk of the ordinance remains in force if any section is declared invalid or unconstitutional. Only the affected section is no longer in force, thus saving the remainder of the ordinance. The pretreatment guidelines ordinance contains examples of several procedural clauses in sections 8, 9, and 10.

In summary, it is virtually impossible to establish and maintain a viable pretreatment program without able legal assistance. The legal authority of the POTW over all of its industrial contributors must be clearly established, the ordinance for industrial use of sewers must be carefully and skillfully drawn, and enforcement must be viewed as a continuing and essential element of the program.

ADMINISTRATION

The scope of any pretreatment program depends to a large degree on a number of factors. Of prime importance is the size of the sewage system and the number and type of industries utilizing the system. Other significant elements include the extent of categorical pretreatment standards requiring enforcement, the type of treatment facility, the water quality standards applicable to the POTW, and the provisions of the plant's NPDES permit.

From an administrative point of view, POTWs have been divided into large, medium, and small systems, referring in general to systems with average dry-weather flows in excess of 100 mgd, between 10 and 100 mgd, and less than 10 mgd. However, a system with a heavy industrial contribution but an average flow of under 10 mgd may opt for a more extensive program, while a system with an average flow in excess of 10 mgd but only limited industrial contribution may adopt a less comprehensive program. Thus, the categorization by size is provided as a general guideline only, and should be considered flexible, depending upon the needs of a particular system. The degree of industrialization in a community, and the type of industrial facilities contributing to the system, are major factors in determining the requirements of a pretreatment program for a given POTW.

The administrative aspects of a pretreatment program relate primarily to organizational and policy considerations, and public relations. Each of these subjects represents a significant factor in the development and operation of a successful program. Provided below are details relating to specific aspects of organization, policy, and public relations which each POTW must consider and resolve in establishing and running its own pretreatment program.

Organizational Structure

A pretreatment program consists of the activities and personnel assigned specific functions and responsibilities in relation to the control of industrial pollutants. For most larger systems, this involves a well-defined organizational structure with assigned personnel having specialized training and qualifications. For smaller systems, it may involve only a part-time assignment for a single individual. However, even those establishing small systems should be aware of the functional steps involved in an industrial pollutant control program and provide for these functions on an appropriate scale.

Figure 3-4 illustrates a conceptual organizational structure based on the functions required for a workable pretreatment program. The manner in which these elements are organized can vary greatly depending on the local situation. The essential aspect is the need for an efficient information transfer mechanism. The typical organization shown outlines the interrelationship of the essential elements of a program.

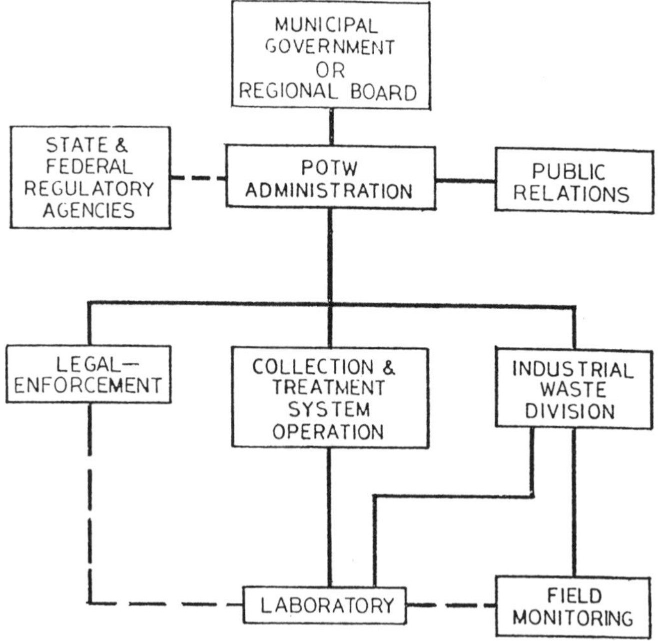

Figure 3-4 / **Conceptual organization of a pretreatment program (4).**

Generally, the larger the system, the more complex the organization. Individual responsibilities also become more clearly defined as the system increases in size. For very small systems, outside experts are frequently utilized to provide engineering, legal, and laboratory expertise. In order to supply the most meaningful information, the organization necessary for an effective pretreatment program is specifically discussed in the following paragraphs in terms of small, medium, and large systems.

Large Systems

Figure 3-5 presents an example of an industrial waste control organization to administer a pretreatment program. The chart illustrates the need for a structured organization to most effectively administer a control program in a large system. The larger and more complex the POTW, the more highly developed and structured the organization should be. Nevertheless, the components of an effective organization demonstrated by the chart are essential for a functional pretreatment program in any large system. An industrial waste control organization for a large system should include the sections detailed below.

Administrative.   Administrative responsibility for the program should rest with a single person who has intimate knowledge of all aspects of pollution control and wastewater treatment within the jurisdictional boundaries. In municipal systems, usually the director of public works or superintendent of sanitation would be the appropriate individual. In regional authorities, day-to-day operations are usually administered by an executive director, general manager, chief engineer, or superintendent. Decisions pertinent to pretreatment policy should be made by this individual drawing upon information supplied by key subordinates. These subordinates should include personnel such as the chief industrial waste engineer, chief plant operator, chief chemist, and other specialists concerned with the control program, including field investigators, engineers, and attorneys. Specific functions of the administration should include: (1) basic policy decisions, (2) management of budgetary needs, (3) personnel administration, and (4) coordination with the public and appropriate municipal, state, and federal authorities.

Industrial Waste Division.   The individuals composing the industrial waste division represent the core of any pretreatment program. Generally, the division is comprised of engineers that conduct the program and field inspectors and laboratory technicians that provide the necessary support for effective operation. The engineering staff engaged in this activity should be the most knowledgeable group of individuals on all aspects of industrial wastewater within a given system. They should be thoroughly familiar with

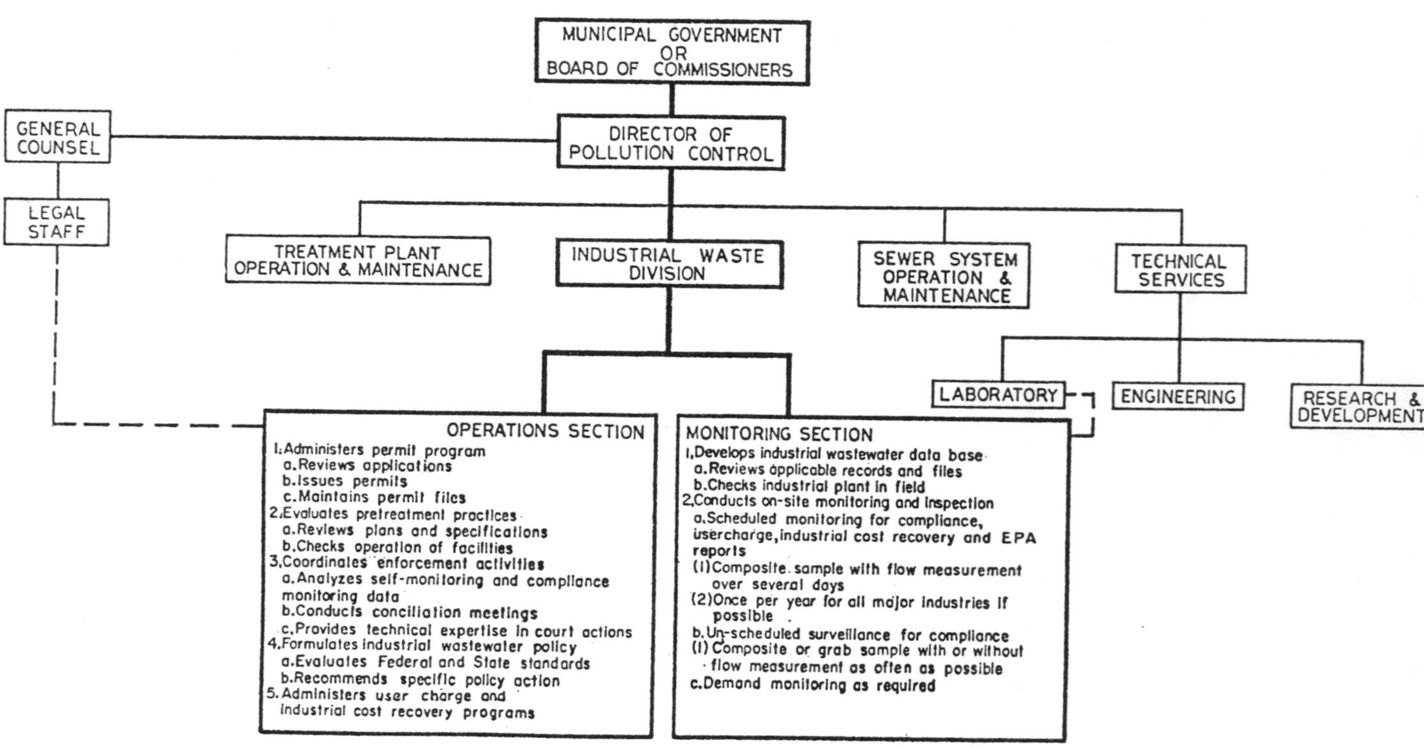

Figure 3-5 / Typical organization of a large system (4).

the operation and wastewater produced by industries in the system, pretreatment facilities utilized by the industries, applicable provisions of federal and state standards and local ordinances, and characteristics of the treatment processes utilized in the system. Generally, this group will formulate industrial wastewater policy and recommend specific policies to the administrator for implementation. The industrial waste division would also generally coordinate enforcement activities and provide necessary technical expertise to the legal staff in court actions.

Staff levels vary considerably depending upon size, fiscal resources, organizational structure, number and type of industries present, and the specific NPDES permit requirements of the system. The largest POTW systems in the country have as many as 50 to 80 individuals reporting to the chief industrial waste engineer. Smaller cities and regional authorities obviously require smaller staffs, with the level generally varying from approximately 0.5 to 2 individuals per 10 mgd, depending upon the factors indicated above.

Field Monitoring.   The organization should include a group of inspectors whose only responsibility is the monitoring of nonresidential contributors. These field investigations should include initial plant surveys, data acquisition at the plant site, and all followup monitoring and inspection activities.

The field monitoring section should have total responsibility for surveillance of industrial sources. Specific functions to be carried out by field inspectors include: (1) sampling and flow measurement at wastewater sources, (2) inspection of plant and pretreatment operations at the time of sampling, (3) maintenance of specialized field equipment, and (4) performance of specialized monitoring activities in connection with locating the source of problems within the system, enforcement activities, etc. The Metropolitan Sanitary District of Greater Chicago provides uniforms and badges for its inspectors to formalize their status in the community. This can be helpful in gaining quick access to industrial facilities when necessary. In any event, all field inspectors should be provided with proper credentials, which should be carried for identification at all times.

Laboratory.   An industrial waste control program will generally require some expansion of the typical laboratory required for control of biological treatment processes. As a result, additional personnel may be necessary to carry out analyses in conjunction with a monitoring program. Frequently, these technicians are incorporated into the existing laboratory organization, so that coordination with the industrial waste division is essential. Some POTWs are structured so that laboratory personnel engaged in industrial wastewater analysis report directly to the chief industrial waste engineer.

Such an arrangement may be preferable for improved communications and delineation of responsibility.

Where this function is part of the overall laboratory responsibility, then data must be reported to the industrial waste division so that pretreatment performance can be evaluated. If any ordinance violations are suspected, then analytical data would also be passed on to the legal staff for evaluation and possible enforcement action. The specific functions of the laboratory in connection with industrial wastewaters include (1) analysis of field samples, (2) maintenance of laboratory equipment, and (3) proper record keeping and reporting in support of industrial waste division activities.

Legal.   One or more attorneys may be required to provide legal services with regard to enforcement of ordinance regulations. Attorneys may not have full-time responsibility in connection with ordinance enforcement. Instead, the legal staff may serve the dual function of supporting legal action against ordinance violators, and general legal support of other activities in the water pollution control area.

For enforcement activity, the legal group should receive information directly from the industrial waste division staff, as well as from field monitoring and laboratory personnel. The special functions of enforcement include (1) assistance in evaluation of suspected ordinance violations, (2) notification of suspected violators, (3) participation in followup meetings with violators (4) preparation of briefs for litigation, and (5) court action.

Medium Size Systems

As the size of a POTW decreases, the operation of the pretreatment program becomes less of a separate entity, and more entwined in the overall operation of the wastewater collection and treatment facility. Because of limited resources, administrative and laboratory personnel generally become responsible for both plant operations and control of pollutants contributed by nonresidential sources. Figure 3-6 provides a typical organization chart for a medium size POTW system.

Since fewer individuals are involved and the organization is not as structured as in large systems, the essential functions of industrial waste control must necessarily be performed in conjunction with other duties. Among these functions is the use of a field monitoring group to carry out plant inspections and effluent sampling. The field crew is essential for all of the specific monitoring requirements necessary to control industrial contributors to the system. Likewise, other specific functions of a pretreatment program would have to be maintained in a medium size system, such as ordinance enforcement, summary and analysis of industrial data, and user charge administration. However, unlike the large-system organization, where

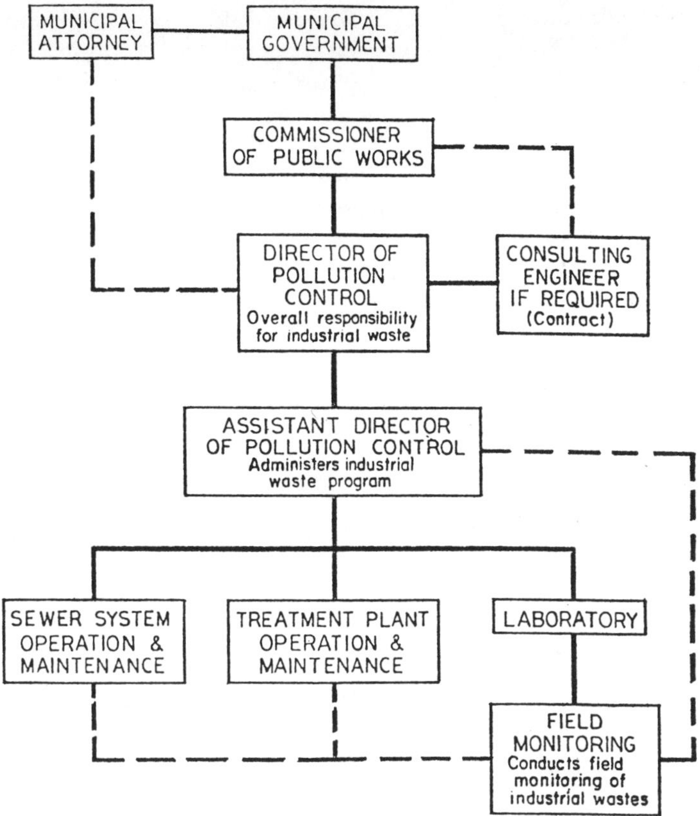

Figure 3–6 / Typical organization of a medium-size system (4).

individuals or groups of individuals have well-defined task assignments, the program for a medium size system will most likely contain individuals with multiple functions and responsibilities. Some medium size systems with many industrial contributors may find it necessary to have an industrial waste engineer or other individual specifically assigned to the control of discharges from these facilities.

Small Systems

A large proportion of POTWs that treat nonresidential wastes fall into the small size system category. In a small size system, adequate resources would generally not be available to provide for one individual whose sole responsibility lies in the area of industrial waste control. Instead, all of the elements of a pretreatment program that are delineated for large systems would have to be handled by personnel currently employed by the municipality. This is not unrealistic, since quite often a small system would be concerned with only a few industries, or even a single industrial contributor.

Figure 3-7 illustrates an organizational arrangement for a typical small system. The structure shown is only one of several that could be effectively utilized in a small municipality. The variety encountered in the organization of small local governments suggests that a number of different arrangements may be equally effective in the administration of a pretreatment program.

In small systems, generally one person has responsibility for monitoring, and all specialized analysis is contracted to commercial laboratories. Similarly, the director of public works would most likely have overall administrative responsibility in addition to performing the functions of the industrial waste engineer. A town engineer may be utilized for industrial waste control, reporting either to the director of public works or directly to the governing

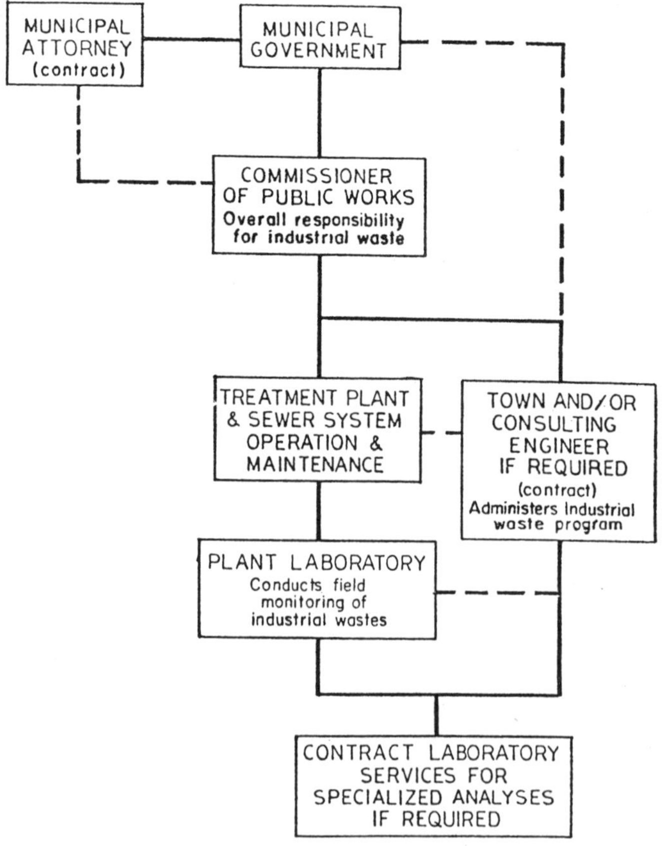

Figure 3-7 / Typical organization of a small system (4).

body. The town engineer may be a full-time employee if the needs of the system dictate, but he is most frequently a consulting engineer under a retainer-type contract to the municipality. Specialized engineering requirements would usually be provided by either the town engineer or another consultant, with legal questions being handled by the municipal attorney.

Policy Considerations

Policy instituted by a POTW must be determined on the basis of conditions within the system. One of the most important factors to consider is the potential effect of the chosen course of action on critical environmental problems such as sludge disposal. As a result, policy should be instituted on the basis of a broad overview of all aspects and consequences of action taken.

Pretreatment

Pretreatment policy is essentially dictated by the federal regulations that have been promulgated to date, including both the general pretreatment standards and the industry-by-industry categorical standards. However, the establishment of a pretreatment program as prescribed in the general pretreatment standards will cause many POTWs to consider the prospect of setting limitations on industrial discharges to their system not covered by federal regulations. In this regard, pretreatment may be necessary for compatible or incompatible pollutants, since both may be limited in the NPDES permit, and since either can cause plant upsets. Generally, however, where design capacity is available, except for shock loading provisions, pretreatment would not be required for compatible pollutants. Pretreatment is most commonly required for incompatible pollutants to prevent interference with treatment processes or passthrough to receiving waters.

Pretreatment of incompatible wastes offers several distinct operational advantages to POTWs. One significant advantage to the municipality is the specialized treatment that each wastewater contribution receives, and the fact that the potential for plant upset is greatly reduced by pretreatment. Local control can also serve as an insurance measure to protect against damage to the POTW from industrial wastes. Finally, the problem of incompatible pollutants contaminating the sludge from the POTW is reduced, thus simplifying sludge handling and disposal operations. This can be a meaningful advantage in terms of the environmental effects of the ultimate disposal of sludge from the POTW. Incompatible pollutants in sludges can cause problems in most disposal techniques utilized, including incineration, landfills, ocean dumping, and land spreading. Consequently, the removal of incompatible pollutants at their source by pretreatment is usually advantageous to the POTW in terms of its sludge disposal.

Nevertheless, incompatible pollutants removed by pretreatment also require ultimate disposition, although the impact on the POTW may have been eliminated. In some cases the sludge produced by pretreatment operations may be pure enough to warrant byproduct recovery or to recycle. When this is not economically or technically feasible, disposal of sludge is necessary. Although the sludges produced by industrial pretreatment may not technically be under municipal regulatory control, the impact on other environmental areas should be noted. A possible approach to this problem would be an effort by the municipality to coordinate off-site disposal with appropriate regulatory agencies. If on-site disposal is utilized by the industry, attempts should be made to evaluate proposed disposal schemes to prevent future air and water pollution problems. A more complete analysis of the impact of pretreatment on sludge disposal is presented in Chapter 5.

Joint Treatment

Joint treatment is a policy alternative that can be potentially advantageous to both the POTW and industry with regard to the construction of new POTW treatment facilities. Generally, the treatment of industrial wastewaters in a POTW is incidental to its primary function of treating domestic sewage. Where the industrial contribution constitutes a significant portion of the total flow and substantially alters the concentration of pollutants normally contained in domestic sewage, the public agency may resort to the joint treatment approach. In this approach, the industry is made a partner in the design and construction of the system, and the treatment works are specifically designed to remove the industrial pollutants. Both capital and operating costs are allocated to the industry and the public agency according to an agreement arrived at through negotiation, or as required by federal regulations if construction grant funds are involved.

Joint treatment of industrial wastewaters with municipal domestic sewage offers these advantages:

—Savings in capital and operating expenses due to the economics of large-scale treatment facilities;
—Increased flow over separate treatment, which can result in reduced ratios of peak to average flows;
—More efficient use of land resources, particularly in cases where available land for treatment facilities is scarce;
—Improved operation (larger plants are potentially better operated than smaller plants);
—Increased number of treatment modules with resultant gains in reliability and flexibility;
—More efficient disposal of sludges resulting from treatment of wastewaters containing pollutants susceptible to treatment in POTWs;

—Utilization of the nutrients available in domestic wastes for biological treatment of industrial wastes which are nutrient deficient.

Possible disadvantages of joint treatment are as follows:

—Where the pollutants are different from those usually treated in a POTW, design to treat the combined industrial-domestic waste stream for these pollutants may not be cost effective.
—Joint treatment by definition implies that the POTW was designed so as not to be interfered with by industrial wastes. However, where this requires design modifications ordinarily not required for domestic wastes, joint treatment may not be cost effective.
—If joint treatment results in sludge disposal or utilization problems it may not be acceptable.
—Some costs for the construction of joint treatment works solely to treat industrial pollutants are not eligible for federal construction grants.

Public Relations

Public relations is an extremely important consideration for any POTW embarking on a pretreatment program. Two distinct aspects must be taken into account: (1) the relationship between the POTW and the public at large, and (2) the relationship between the POTW and its industrial contributors.

In general terms, the POTW should strive to keep the public informed about its programs and policies, and attempt to get the community involved and behind the pretreatment program. This can be done in a variety of ways, but typically public meetings and meetings with specific public interest groups can prove to be very helpful. In addition, newspaper, radio, and television advertising, although costly, reach large numbers of people and are extremely effective in disseminating information. To the extent that the resources of the POTW permit, the use of newspapers, radio, and television should be strongly considered to gain public understanding and support for the pretreatment program.

The general pretreatment regulations contain specific requirements with regard to public participation. POTWs are required to comply with the stipulations of 40 CFR, Part 25 (public participation). These regulations provide guidance for public participation relative to informational materials, consultation, access to information, enforcement, legal proceedings, rule making, reporting, and public meetings. POTWs instituting a pretreatment program must become familiar with these regulations. In addition, POTWs are required to publish on an annual basis a list of all industries in the system that have significantly violated the pretreatment requirements at least once, and a summary of the enforcement actions taken by the POTW. This notice must be printed in the largest daily newspaper published in the municipality in which the POTW is located.

With regard to the industries discharging into the POTW system, the thrust should be to develop a spirit of cooperation in achieving the goals of the pretreatment program. It must be recognized that some industries using a public sewage system will be reluctant to provide the necessary pretreatment facilities, particularly if they have been using the public system for some time and the effects of their contribution are not apparent. It is incumbent upon the POTW when instituting a pretreatment program to attempt to develop a feeling of mutual trust between POTW and industry representatives. This can be best accomplished through a series of meetings with industry personnel to inform them of the regulations and programs involved and to provide a forum for industry to present their suggestions, concerns, and recommendations.

One way in which some public agencies have been able to enlist the support of their contributing industries is through the formation of an informal coordinating committee, which would consist of representatives of the affected industries and the public agency. This committee would explore all aspects of the federal regulation, including pretreatment requirements and the NPDES permit for the POTW, and develop programs which would meet the requirement of the regulatory agencies. Such committees would be purely advisory and would not have any legal status, but could serve as a forum for the exchange of ideas.

In any case, public relations is a subject that is often overlooked, frequently with significant ramifications. The need to involve the public at large is specified in the regulations; however, the spirit in which a public participation program is instituted is extremely important. Public participation should be viewed as an opportunity to generate community support rather than as an obligation. Similarly, the approach taken with industrial contributors should foster cooperation and mutual trust instead of leading to adversary roles and polarization.

MONITORING AND REPORTING

Elements of an Industrial Monitoring Program

The management of an industrial waste control program within a POTW system requires a constant flow of information on the quality and quantity of industrial contributions to the publicly owned system. In general, the function of a monitoring program is to provide a mechanism by which the POTW operator can obtain information on the pollutants introduced into the sewer system. The information obtained through monitoring activities may then be applied to specific areas of concern to the municipality. These specific areas include compliance with ordinance requirements, ascertaining user charge fees, and completion of reports required by EPA. Additionally, the POTW operator may use monitoring information to determine the contributors who are responsible for releasing materials potentially harmful to collection or treatment systems.

Depending on the specific situation, information obtained by monitoring may also be used in the development of the ordinance and in its enforcement. Monitoring information can be especially useful in developing those sections of the ordinance that set levels for incompatible pollutants, as well as determining orders of magnitude for an equitable system of user charge fees. As a result, ordinance development and enforcement work hand in hand with monitoring activities.

Although monitoring in a broad sense performs the single function of obtaining quantitative and qualitative information on nonresidential contributors, specific subfunctions should also be completed to provide a total program. For monitoring, there are well-defined intermediate steps that should be accomplished during the course of the overall program. In a well-managed system for industrial pollutant control, information should be transferred in a closed loop, where monitoring, ordinance compliance, ordinance enforcement, and user charge fee determinations all input to one another.

The extent of monitoring programs instituted by POTWs are dependent on the resources, both financial and technical, of the municipality or authority. The need for a detailed program depends largely on the prevailing local conditions. Factors such as size, amount of industrial flow, and nature of contributing industries have a large impact on the magnitude of required monitoring programs. Especially important are the prevalence of toxic pollutants in industrial contributions as these materials come under increasingly more stringent control. No matter what the specific situation, all POTWs that institute an industrial pretreatment program will need minimum information on the quantity and quality of wastewaters contributed by nonresidential sources into the sewage collection system. As such, monitoring, i.e., the process of systematically sampling and analyzing industrial contributions, is an essential part of any pretreatment program. The specific purposes of monitoring efforts within the context of an overall industrial waste control program are outlined below.

Purposes

Reports to Permitting Authority

As part of many NPDES permits, POTWs are required to provide the NPDES permitting authority, which is either the EPA regional office or a designated state agency, periodic reports concerning the nature of industrial contributions to the system. An industrial monitoring program is usually the means by which the information required for these reports is collected.

Compliance with Pretreatment Standards

Pretreatment standards fall into three broad categories: general prohibitions, national technology-based standards, and locally developed standards. Under the revised federal pretreatment regulations (1), POTWs will be required to enforce the first two of these categories of pretreatment standards. Additionally, under the new regulations, local programs will required which may result in the development of local pretreatment standards. In either case, monitoring of the industrial contributions to a POTW will be required to assure compliance.

Ordinance Development

An ordinance will generally prohibit, or set limits on the release of specific materials to the municipal wastewater treatment system. Pollutants other than those covered by federal pretreatment standards that may either inhibit biological processes, or pass through the system, may be limited by ordinance. It is therefore essential that the monitoring program supply data on the contributions of prohibited and limited substances. These data are essential to development of the ordinance limits as well as to the ongoing enforcement of ordinance requirements.

Analytical Data Base Development

Control of pollutants contributed to POTW systems by industries requires a thorough knowledge of the normal concentrations of pollutants in the collection and treatment system. Information on background levels can shed light on the proportion of a particular pollutant that originates at industrial facilities and is therefore controllable. Mass balance calculations for the total system can provide information on the impact of dilution. A well-conceived monitoring program can form the data base necessary to quantify both background concentration and dilution characteristics for the system.

Enforcement

An important function of a monitoring program is its ability to provide specific information required by enforcement activities. Enforcement implies that there has been a violation of a regulation. In general, municipalities will use monitoring information to assess deviations from ordinance stipulations. If conducted with the proper rigor and quality control, wastewater sampling and analysis previously performed by the municipality can be used in enforcement activities directed at ordinance violators. The final product is defensible data, which when applied to compliance with the ordinance or other standards, provide the link between monitoring and the overall control program.

Types of Monitoring

A systematic approach to a pretreatment program yields four major types of monitoring activity. These four monitoring methods are aimed at obtaining related information on industrial wastewater characteristics tailored to serve specific data needs. Although only two of the types of monitoring outlined below are specifically required by federal regulations, a complete program will need the kind of information supplied by all four monitoring activities. Since monitoring yields the continuing data necessary to administer the pretreatment program, all efforts should be made to institute all four monitoring modes.

Scheduled Monitoring

Scheduled monitoring involves the systematic sampling and inspection of significant industrial contributors to the POTW system in accordance with a predetermined schedule. The schedule should be developed by the POTW personnel administering the industrial pollutant control program, and should be maintained as confidential information so that the industries in the system remain unaware of contemplated monitoring visits in advance of their occurrence. Notification is necessary, however, prior to the visit when the sampling point is located on plant property, to arrange for required utilities and proper access to sampling areas. The schedule should attempt to provide for the monitoring of each significant industrial contributor at least once per year if at all possible. If resources do not permit this type of coverage, then visits should be planned as often as possible within the limitations of available resources.

Scheduled monitoring should serve a number of specific needs, including: (1) checking for compliance with the ordinance, (2) user charge and industrial cost recovery determinations, and (3) completion of required EPA reports. Scheduled compliance monitoring should be aimed at obtaining all the information necessary to determine adherence to the local ordinance. The information required for user charge and capital cost recovery coincides to a large degree with data necessary for the completion of required EPA reports.

In general, scheduled monitoring would include on-site inspection of pretreatment facilities and plant operations, and composite samples and flow measurements taken over a period of several days. On-site inspection is necessary to insure that pretreatment facilities are being operated properly and to detect any dilution of plant wastewaters. The determination of flow is important for several reasons. Primarily, flow information is essential in the calculation of user charge fees. Additionally, flow readings are necessary as an added safeguard against dilution and to confirm the validity of concentration measurements taken for ordinance compliance.

The degree of care and conformance to established procedures in obtaining samples and flow measurements during scheduled monitoring visits is an extremely important consideration. Information obtained during this process may ultimately be used in enforcement activities culminating in court action. Additionally, in the determination of user charges, the municipality may be compelled to monitor with sufficient rigor to satisfy the industry that the information obtained is sufficiently accurate. This situation is especially valid in those cases where sewer use charges are a significant expense for the industry.

Unscheduled Surveillance

In addition to the planned approach, POTWs should institute a less formal type of compliance monitoring designed to provide a spot check of industrial contributors. This random type of compliance monitoring should be focused on maintaining a degree of surveillance, and would generally not be formally planned or announced by the POTW. The general federal pretreatment regulations (1) specifically require the POTW to conduct this type of random sampling of industrial contributions to the sewer system. Additionally, data obtained from random sampling must be kept on file and made available on request to EPA.

Scheduled compliance monitoring is designed primarily to establish the characteristics of contributions from major sources, while unscheduled surveillance attempts to survey all sources within the system at random. By checking all contributors over an extended period of time, the municipality can continue to expand its data base, and keep abreast of trends and changes within the system. This type of monitoring must be conducted on a random basis, with contributors being observed during normal operation, thus providing information on the true nature of the wastewater. If ordinance violations are suspected, the information obtained during unscheduled surveillance can also be used to evaluate the need for further, more detailed, evaluation of a particular contributor.

Unscheduled surveillance can be conducted with less rigor than scheduled compliance monitoring. Since surveillance activities are intended only to provide a quick spot check to determine the need for future more detailed analysis and evaluation, this type of monitoring can be less formal than a full compliance monitoring visit. Unscheduled surveillance should involve no more than a few samples, perhaps composited over a short period of time, and a cursory inspection of plant operations and pretreatment activities. Where resources are limited, grab samples are frequently used for this type of surveillance. Flow measurements should be taken if facilities are available to obtain readings without difficulty.

Demand Monitoring

As the name implies, demand monitoring should be conducted when an upset or other disruption of system operation occurs, which may have been caused by an industrial source. Additionally, any discharge of prohibited or limited materials can prompt demand monitoring. The intent of demand monitoring is to locate the source of prohibited materials in the collection system. However, locating the source is often difficult and as a consequence demand monitoring frequently fails to pinpoint the origin of the pollutants in question. Nevertheless, demand monitoring should be conducted in a visible fashion. Consistent efforts to track down violators are not only designed to catch the offender, but to eliminate future similar discharges of prohibited materials by projecting an image of quick response to violations or emergency conditions. As a result, demand monitoring should be initiated rapidly and in a conspicuous manner to assure public awareness. Specific occurrences or emergency conditions that may initiate a demand monitoring effort are detailed below.

Contributions of Explosive or Corrosive Material to the Sewer.   Release of these types of materials are generally prohibited by the municipal ordinance. Because of the magnitude and immediacy of the potential impact of explosive or corrosive materials, swift location of the source is essential. The most effective means of locating sources of explosive or corrosive materials is by utilizing sewer backtracking, which is a systematic search upstream through the sewer system until the source is pinpointed. It should be noted that whenever sewer backtracking is initiated, special attention should be paid to safety considerations. This is especially true when the suspected violation involves explosive materials.

Operating Difficulties.   Treatment plant and collection system operating difficulties can also prompt demand monitoring. One of the more serious operating problems is caused by the release of materials into sewers which can cause blockages or plugging. Similarly, excessive quantities of viscous or floating solids entering a treatment plant can disrupt unit operations such as sludge digestion. The presence of excessive foaming can also cause operating difficulties, and may prompt a demand search. In general, any upset of normal operating routine may be considered cause for initiating demand monitoring.

Violation of the POTW's Permit Requirements.   A POTW's NPDES permit generally contains limits on the concentrations of specific pollutants that it can discharge to navigable waters. If the treatment process is disrupted to the extent that these limits are exceeded, it becomes the responsibility of the municipality to determine the source of the materials that might be passing through the system, or causing poor removal of the pollutants that the treatment system is designed to remove.

Violation of Pretreatment Regulations.   Since NPDES permits will now require the development of a control program that contains requirements for compliance with federally mandated and locally enforced pretreatment regulations, a suspected violation of these standards can initiate demand monitoring. In such cases, demand monitoring should provide information on the cause of interferences, and the responsible party. In these situations, demand monitoring can be most successful when the municipality has access to a good data base. Using data base information, the probable sources of interfering materials can generally be determined and monitored so that responsibility can be properly assigned.

Self-Monitoring

Many of the monitoring activities that have been outlined in the preceding paragraphs require that samples be taken at the effluent of an industrial contributor, and analyzed for appropriate pollutant parameters. Depending on the available resources and manpower, the municipality may not be able to perform all of the various monitoring functions required for industrial contributors. At the minimum, the municipality should attempt to conduct, with its own personnel, some sampling at all significant industrial contributors within the system. However, complete coverage of all contributors to the POTW may be more difficult to implement using municipal personnel and resources. One way to circumvent this problem is to require each major contributor to do its own sampling and analysis, a function which is usually termed self-monitoring. Federal regulations require that all industrial sites for which national categorical pretreatment regulations have been promulgated must initiate self-monitoring 90 days after the effective date of the regulations and twice yearly thereafter. The results of these self-monitoring activities are then to be kept on file and made available to EPA on request. Although the federal self-monitoring requirements are only required for industries covered by national categorical pretreatment standards, the POTW may elect to require additional industries within the system to conduct self-monitoring.

Self-monitoring, although cost effective for the municipality, does have some serious drawbacks. First, the objectivity of data generated by industrial self-monitoring may come into question. The industry may elect to temporarily modify plant operations, or sample in such a fashion that the pollutant load from their plant is minimized. Additionally, the use of self-monitoring information in an enforcement procedure is questionable. The use of data provided by an industry in enforcement activities against that industry raises serious legal questions. In any event, self-monitoring data can serve as an important supplemental source of data. Self-monitoring systems require that a mechanism for reporting and record-keeping be maintained by the industrial establishment. In addition to maintenance of a data file by the industry, periodic reports should generally be sent directly to the POTW. The record-keeping function permits access to a history of source quantity and quality which can aid in both data base development and compliance determination.

Field Considerations

Background

Organizational and managerial aspects of a monitoring program may vary considerably from system to system, but the approach to the technical problems encountered in any field monitoring activities remains fairly constant. The need to maintain rigor and objectivity dictates that sound, uniform and well defined procedures be maintained during plant investigations and sampling programs. Some guidance on how to develop and carry out a monitoring program is available in an EPA technology transfer document entitled *Handbook for Monitoring Industrial Wastewater* (7). This document describes technical aspects of monitoring, but its major emphasis is directed at industries that are discharging directly to navigable waters and are engaged in self-monitoring activities. Although many of the details remain the same, field considerations for monitoring industrial contributors in a pretreatment situation have a slightly different orientation. Special field considerations for monitoring pollutants introduced into POTWs are outlined in this section.

Preparation for a Monitoring Visit

Industries should not be notified well in advance of scheduled monitoring or unscheduled surveillance visits. It is generally agreed that by not giving prior notice to the industry, the samples that are obtained will be more representative of daily operation. Nevertheless, notification just prior to the visit is necessary when the sampling point is located on plant property, to allow for necessary utility and access arrangements by plant personnel.

Prior to sampling at a contributor, the sampling crew or inspector should obtain specific information about the industry. During the initial visit, a plant inspection report should be prepared.

Several specific items should be included in this report:

—A sketch of the location of all wastewater effluent lines that flow into the publicly owned sewer system (the sketch should also include the layout of major plant features);
—A description of major product lines and processes utilized within the plant;
—A detailed description and appropriate sketches of existing pretreatment facilities, including operating data if available;
—A list of pollutants of interest at the plant, with emphasis on materials limited or prohibited by the ordinance;
—A list of raw materials or products stored at the plant that may have a pollutant potential.

A procedure adopted by some POTWs with established industrial wastewater control programs has been to formulate a questionnaire form covering, at the minimum, the items listed above. The form would then be sent to the industry well in advance of the inspection visit so that the industry would have an opportunity to review it. When this procedure is followed, plant personnel will be able to easily provide the information required by the form. The form would then become part of a permanent file on the industrial contributor. Every time future sampling would be conducted at the industrial plant, the file would be reviewed and updated, providing a record of progress toward compliance and of changes in plant processes.

Scheduling

An attempt should be made to systematically cover all significant contributing industries annually if resources permit. After initial data base development, monitoring should be conducted at each major contributor to determine progress toward compliance. Once compliance is achieved, the contributor must be sampled periodically to assess continued compliance. Additionally, when a scheduled compliance monitoring visit uncovers a violation, scheduling should be altered to provide more detailed analysis of the wastewater. This extended sampling program is normally required to obtain data sufficiently sound for use in enforcement activities.

In all cases, the frequency and order of investigations should be determined on the basis of size and importance of the contributor. Sampling should be started with the largest, or most significant industry. Once the major industries are covered for data base and initial compliance purposes, a continuing program of followup monitoring should be instituted. It may not be within the resources of the municipality to cover all contributors within its system thoroughly and equally. A schedule in which major contributors are monitored more frequently than minor contributors will generally be necessary.

Sampling and Flow Measurement

Accurate flow measurement and sampling techniques are of prime importance in obtaining valid monitoring information. Both flow measurements and sampling can be accomplished either manually or through the use of automatic devices. Three types of sampling may be utilized:

1. *Grab samples.* In this type of sample, a single volume of wastewater is obtained and analyzed. This procedure will not always provide an accurate measure of wastewater characteristics, especially when the flow is heterogeneous, or varies with time.
2. *Simple composite sampling.* This is a timed, sequential collection of equal-volume grab samples that are combined in a single reservoir. This type of sample can give a partial evaluation of the variability of wastewater composition with time. It does not provide any measure of the total pounds of pollutant discharged, since pollutant loading is a flow related value.
3. *Flow-proportioned composite samples.* These are obtained by collecting incremental samples with volumes proportional to flow. This type of sample, when analyzed and compared to total flow, provides the most accurate measure of wastewater quality and pollutant loading.

Automatic sampling devices that can obtain samples of all three types are commercially available. These automatic samplers vary over a wide range in cost, applicability, and reliability. Two EPA documents are available which provide thorough evaluation of commercial automatic samplers;

—*Sampling of Wastewater,* by Philip E. Shelley (8);
—*Wastewater Sampling Methodologies and Flow Measurement Techniques,* by Daniel J. Harris and William J. Keffer (9).

The second document also includes information on the performance of portable automatic flow-measuring devices. There are considerably fewer devices of this type on the market than there are automatic samplers. However, a few portable instruments are available that can provide reasonably accurate flow measurement data.

For the smaller system, or those municipalities that have only limited resources, both sampling and flow measurement may be accomplished using manual techniques. Samples of all three types can be collected using manual methods. In general, there is little equipment commercially available for manual sampling. What is usually needed is an extendable pole, with a stoppered bottle attached to the end. This type of sampling pole can be easily fabricated. The sample bottle should be hinged so that it can be tilted to align parallel to the wastewater flow. This orientation allows for sampling from very shallow streams. The bottle stopper would be attached to a string so that it can be removed while the sample bottle is submerged.

Magnitude of Errors

An often overlooked factor in wastewater evaluation is the magnitude of errors that may occur during data collection in the field. Special attention should be paid to the accuracy of sample collection activities. Sampling errors can range up to 200% of the true value. This is because typical industrial waste may have a large proportion of its pollutants in the form of suspended solids. As a result, it is important that the quantity of suspended solids entrained during sampling be proportional to the suspended solids content of the total wastewater stream. Common practice is to simply place a weighted suction tube in the wastewater flow, or to immerse an open sampling bottle in the stream. Since solids entrainment is a velocity-controlled process, an attempt should be made to obtain samples isokinetically. Ideally, during sampling there should be a minimum velocity difference between the interior and exterior of the suction tube. In practice, accomplishing this type of sampling is a difficult procedure, but the situation can be significantly improved by aligning the sample tube so that it is facing upstream and is secured rigidly in place. Additionally, a sufficiently high velocity in the sample tube should be maintained to overcome gravity settling in vertical tube runs and to provide a degree of scouring to avoid solids settlement. EPA recommends a velocity of 2 feet per second (fps) in the sample tube.

It should not be inferred from the foregoing that perfectly accurate sampling can be accomplished by implementing the techniques outlined. Wastewaters are generally heterogeneous and rarely flow at a constant rate for any extended period of time. However, by establishing a uniform methodology for sampler suction-tube placement, samples taken at different times and even different locations will have the overall sampling bias decreased. This is not to say that all sampling errors will be eliminated, but at a minimum uniform sampling techniques will provide a certain degree of added comparability between samples.

With regard to flow determination, the accuracy of any flow measurement depends greatly on the control surface utilized. Some ordinances may require that major contributors install a special control manhole designed to provide sufficient access for sampling and an appropriate control surface for flow measurement. Permanent flow recording equipment may also be required.

The installation of a standard weir or flume makes flow measurement a simple matter of measuring wastewater depth. No control surface is completely accurate, but the combined use of a quality automatic flow measuring device and a control surface can typically yield flow measurement accuracy of better than $\pm 15\%$.

Sample Handling

Once an accurate sample has been obtained, several precautions should be taken to assure that the validity, objectivity, and defensibility of the monitoring operation is maintained. Because the analytical values resulting from monitoring activities may be used in enforcement proceedings, well-established, standard sample-handling procedures must be instituted. These procedures fall into two broad categories: technical and administrative.

Sample preservation is a technical consideration which is dependent on the specific parameters that will be measured in a wastewater sample. In general, the pollutants in industrial contributions to municipal collection systems of primary interest to the POTW are the so-called classical pollutants (BOD, COD, suspended solids, etc.) and the 129 toxic (or priority) pollutants established by EPA for limitation in the effluent from 21 major industries. Within these two groups of analytical parameters there are various subgroups, or fractions, which require differing procedures for preservation at the sampling point. The purpose of the preservation techniques is to halt any chemical or biological processes that may be occurring in the sample which might alter the concentration or form of the pollutant in question.

For the purposes of sample preservation and the subsequent analysis, the classical and priority pollutants may be subdivided into several major groups:

—Classical pollutant parameters;
—Semivolatile organic compounds;
—Volatile organic compounds;
—Metals;
—Pesticides;
—Phenols (total);
—Cyanides;
—Asbestos.

Each pollutant group requires specific sample-handling and preservation techniques. Actual sample preservation techniques are outlined in various analytical handbooks such as *Standard Methods* and the EPA *Chemical*

Methods Manual. The content of these and other similar handbooks are discussed in more detail in the section devoted to laboratory considerations in monitoring.

In general, preservation techniques may range from simple cooling to 4°C or addition of stabilizing chemicals to complex procedures associated with samples taken for priority pollutant analysis. Because of the heterogeneous nature of many industrial wastewaters, the parameters to be analyzed may require the use of incompatible stabilizing reagents. To solve this problem, it is usually necessary to take a relatively large volume of sample, so that it may be divided for appropriate preservation. For analysis of the organic priority pollutant fractions, samples must be collected in specially prepared, organic-free glassware.

Another technical area of concern is the possibility of contamination at the time of sampling. The range of expected concentrations of classical pollutants is high (milligrams per liter and up) relative to typical occurrences of priority pollutants (in the microgram-per-liter range). As a result, contamination can result from unexpected sources. A typical example is sampling for subsequent analyses for phthalate esters, a group of priority pollutants. Automatic samplers may be used, but a certain amount of phthalates from the plastic tubing used in the mechanism can be expected to leach into the sample. Specialized samplers are available that substitute glass and teflon for plastic parts, eliminating the problem. Alternatively, organic-free water may be passed through the sampler and analyzed, providing background values. This example illustrates the potential for contamination, especially during priority pollutant sampling. In general, there are many complications in looking for priority pollutants in industrial contributions. EPA has issued several protocols for priority pollutant sampling and analysis which should be studied in depth before detection of these materials is attempted. These protocols are discussed below in the section on laboratory consideration.

Once an appropriate sample is obtained and properly stabilized, it is essential that the possession of the sample be properly documented. That is, the person completing the field sampling should maintain a log containing pertinent information such as date, time, and location of the sampling activity. Before releasing the sample to the laboratory, or any other appropriate official, a signed receipt should be obtained documenting the exchange. As the sample is transported, a continuous history of its condition and locations should be maintained through successive log entries and receipts.

This chain of possession or chain of custody is a vital administrative procedure. The reason for such caution in the handling and transfer of samples stems from the need to be certain of sample integrity as part of any enforcement activity. It should be assumed that every scheduled compliance monitoring sample may become evidence in a court of law. In practice, few ordinance violations will require legal action, but sample integrity must be maintained nevertheless. If the municipality cannot prove that a sample has not been mishandled or tampered with, then any inferences regarding the quality of the wastewater that the sample represents fall into jeopardy. It is therefore essential that a chain of possession be maintained and recorded.

A final administrative sample-handling consideration relates to the situation in which the facility being sampled requests parallel sample portions for separate confirming analysis. These samples, commonly referred to as split samples, are usually obtained by dividing the collected samples and providing half to the plant. Split sampling may be required at the discretion of the plant being sampled. In all cases, the parallel analysis can provide valuable backup information, as well as fostering goodwill with the industry being sampled.

On-Site Safety

Visitors to industrial establishments are usually required to abide by any safety regulations observed by plant management. Ideally, access to wastewater flows should be available from manholes or junction boxes located outside of plant property. In this way, inspectors would not be exposed to any hazardous operating activities and would be able to sample without the need to gain entry to the plant. This eliminates any complications, including possible liability entanglements.

Unfortunately, accessible manholes located outside of plant boundaries are rare, and consequently inspectors are generally required to enter plant property. Under these circumstances, all pertinent safety rules in force at the plant must be adhered to. The first step upon presenting credentials and entering the plant, is to notify appropriate plant management, and request applicable safety information. In most cases, a formal set of safety rules are in effect, and these rules should be recorded and filed with the plant inspection report for future reference.

Although most POTW administrative personnel are well aware of safety precautions associated with typical sewer maintenance activities, such as traffic control in the area of street crew operations and the hazards of manhole entry, these practical safeguards may not be familiar to the personnel carrying out industrial monitoring activities. In this regard, standard safety steps and use of typical safety equipment should be practiced by monitoring teams. Of utmost importance are the precautions associated with entering manholes, especially with regard to testing for hazardous atmospheres. Standard gas test kits, breathing apparatus, sewer ventilating equipment, and safety harnesses must be used. Since sewer entry may become a routine part of the monitoring program, expense should never become a deterrent to use of appropriate safety equipment and procedures.

Equipment Setup and Field Analysis

In the absence of a control manhole equipped with brackets or ledges for mounting automatic samplers and flow measuring devices, a method must be devised for securing and mounting equipment. Within the boundaries of a plant some degree of security can be assumed, but unattended sampling points in public areas can be tempting to vandals. One solution to this type of field problem is the use of a self-contained trailer or van outfitted with appropriate materials and equipment. A further advantage to using a monitoring vehicle is that the time required for setup and removal of equipment is greatly reduced. Furthermore, the vehicle can be outfitted with appropriate field analytical equipment. Certain parameters require immediate analysis upon sampling, such as dissolved oxygen and pH. Equipment for these analyses can be mounted in the vehicle to facilitate rapid analysis.

For the small system, resources may not be available, nor the need apparent, for a sophisticated, well-equipped monitoring vehicle. At the very least, a compact package containing necessary equipment should be utilized. A minimum number of separate equipment packages should be maintained, with the dimensions of a foot locker being an ideal size for each equipment carrier.

Use of Control Manholes

For the larger contributors, the frequency of sampling activities may make the use of a permanent sampling facility cost effective. A manhole through which the total plant sewer discharge flows is commonly outfitted with permanent equipment for sampling and flow measurement. At the minimum, brackets and ledges appropriate for mounting flow-measurement and sampling equipment should be provided. More sophisticated devices, such as flow proportioning and continuous analytical recorders (pH, dissolved oxygen, etc.) may also be installed. The complexity of the required control must be judged on the basis of the importance of the contributor within the POTW collection system. A complex sampling or control manhole may not be necessary for a minor contributor, whereas complex arrangements of flow-measuring devices, sensors, probes, and recorders may be needed at the larger industrial sites. In any event, the nature and scope of the required facilities should be firmly set down in the municipal industrial waste control ordinance.

Continuous On-Line Monitoring

In recent years, equipment has been developed which can automatically monitor various wastewater characteristics. For industrial wastewaters, numerous constituents may be of interest. However, as a result of operating difficulties, generally only a few parameters can be successfully analyzed on a continuous basis.

The sensors typically used in automatic monitoring equipment are often very sensitive to interferences found in wastewater, and as a consequence great care should be exercised in choosing this type of specialized equipment. Because of the commonly encountered operating difficulties, continuous on-line monitors generally require a high level of maintenance and attention. If the equipment works well, it can serve as an excellent source of continuous data which will not only aid in wastewater and process operation evaluations, but through the use of auto-punching can also reduce the amount of manual handling of data.

Various manufacturers offer continuous on-line analytical equipment which can provide excellent monitoring information. However, it should be noted that automatic analysis is still a developing technology, which should be approached with an appropriate degree of caution.

Laboratory Considerations

Background

Once an administrative approach and technical methodology are developed for obtaining industrial wastewater samples, a mechanism for accurate and rapid analysis of the samples must be provided. It is essential that analytical results be accurate and reproducible to ensure that monitoring activities will provide the quality of information necessary for a successful pretreatment program, not only from the standpoint of providing fair assessment of industrial contributions, but to provide defensible data that may be required for enforcement activities.

Standard Analytical Techniques

Precise and well-recognized techniques have been established for the analysis of wastewaters. EPA has promulgated rules and regulations on this subject entitled *Guidelines Establishing Test Procedures for Analysis of Pollutants—Proposed Regulations,* dated December 3, 1979. (10) These rules stipulate specific analytical methods that are recommended by EPA for the determination of 228 chemical and biological parameters many of which are as reported in three analytical handbooks. The three referenced manuals are:

—*Manual of Methods for Chemical Analysis of Water and Wastes,* available through EPA technology transfer (11);
—*Standard Methods for the Examination of Water and Wastewater,* published by the American Public Health Association (12);

—*Annual Book of Standards, Part 31, Water, Atmospheric Analysis, 1975,* published by the American Society for Testing and Materials (ASTM) (13).

Each of these documents provides a short synopsis of the analytical method for each parameter, information on interfering substances, and step-by-step instructions on how to carry out the analysis. Also included is information on the calculation of results, the precision and accuracy of the analytical method, and techniques for chemically stabilizing and preserving samples.

These three publications should be considered a minimum requirement for any analytical laboratory. They supply the basic information that a trained lab technician would need to successfully perform nearly all analytical procedures for traditional parameters that may be required in connection with monitoring programs.

Priority Pollutant Analysis

In contrast to the well-established procedures for analytical evaluation of traditional or classical pollutants in wastewaters, analysis for the organic fractions of the priority pollutants is a currently developing science. Whereas the metallic priority pollutants (copper, chromium, lead, etc.) may be analyzed by standard wet chemical techniques or atomic absorption methods, both of which are well documented in the standard analytical manuals of methods, the organic priority pollutants, which are often present in the minute microgram-per-liter range, may require complex analytical protocols leading to gas chromatography–mass spectrometry (GC–MS) analysis. In order to effectively use GC–MS, various preliminary separation techniques are required. The volatile organic fractions require a helium purging technique in which the volatile constituents in a wastewater are absorbed onto a sorbent trap with subsequent desorption prior to GC–MS evaluation. The semivolatile fractions require various extraction steps utilizing either acidic or basic conditions to separate groups of organic priority pollutants for subsequent analysis. These separation techniques require a high level of expertise which is usually beyond the scope of even the moderately sized POTW.

For those wishing more detailed information on the analytical techniques involved in organic priority pollutant analysis, reference should be made to the various protocols for analysis developed by the EPA effluent guidelines division. Specifically two documents should be referred to:

—*Sampling and Analysis Procedures for Screening of Industrial Effluents for Priority Pollutants* (14);
—*Analytical Methods for the Verification Phase of the BAT Review* (15).

These documents were developed for the specific purpose of providing a set of analytical methods to be used by the EPA effluent guidelines division in evaluating industrial discharges for subsequent limitation of priority pollutant concentrations in their effluents. They were not devised to be used in general for routine analysis. The proposed regulations promulgated by EPA (10) do, however, set forth specific procedures for all 129 priority pollutants.

Many concerned with pretreatment, especially for industries known to have priority pollutants in their discharges, may have the impression that these complex procedures may be needed, or that analysis for all 129 priority pollutants in a wastewater sample may be required by a regulatory agency. In reality, it is likely that only a few priority pollutants will be of importance in any particular category of industrial discharge to a sewer system. Further, the current policy at EPA is to encourage the use of so-called surrogate or indicator parameters for priority pollutants which are difficult or expensive to analyze. As a result, a correlation between, for example, total organic carbon (TOC) and an exotic priority pollutant may be provided such that TOC analysis may be substituted for complex GC–MS techniques. In any event, very large POTWs may have a need for their own GC–MS capabilities, in which case the protocols cited above would be applicable.

Analytical Quality Control

The potential errors encountered during analysis of wastewater samples, although not as great as the errors associated with poor sampling and flow measurement techniques, can, nevertheless, have a great impact on the acceptability of monitoring information. Without the aid of independent checks and general quality control, the lab technician can report erroneous results without being aware that a problem exists.

Analytical quality control assistance is available in several forms from EPA. A document entitled *Handbook for Analytical Quality Control in Water and Wastewater Laboratories* (16) has been published by the EPA technology transfer program. In this handbook, specific information is provided that can guide the lab technician or chemist toward sound and reliable techniques and procedures. In addition, standard approaches to data handling and reporting, and information on calibrating equipment are included.

The EPA *Manual of Methods for Chemical Analysis of Water and Wastes* (11) also includes information pertinent to laboratory quality control. Each of the ten EPA regional offices has an office of analytical quality control headed by a regional analytical quality control coordinator. Included in the EPA methods manual are a listing of the ten quality control coordinators, accompanied by appropriate addresses and telephone numbers. Through the quality control coordinators, any interested party can obtain preanalyzed samples that can be used to test the accuracy of analytical techniques. Periodic analysis of known samples can provide the lab technician with an independent check of his accuracy, providing the opportunity to correct any improper procedures.

In general, special techniques such as replicate analysis, spiked samples, and frequent checks on equipment calibrations are essential parts of any laboratory quality control methodology. The bottom line is reproducability of data and its subsequent integrity in enforcement proceedings. As such, special attention must be paid to quality control activities.

Laboratory Equipment

Laboratory equipment required for a municipal wastewater treatment plant is generally standard, with variations occurring only in degree, as the size of the plant varies. If, however, a publicly owned system receives wastewater from a particular industry, specialized analytical equipment may be required. The range of applicability of several special types of equipment are discussed in the following paragraphs.

Atomic Absorption and Emission.   Atomic absorption spectrophotometry provides a rapid and easily performed technique for the analysis of metals in wastewater. Nearly all of the more than thirty elements that can be analyzed by atomic absorption can be analyzed by standard wet chemical techniques. However, the wet chemical methods can be tedious and time consuming, requiring detailed sample pretreatment procedures. Atomic absorption methods provide for metal analyses with a mimimum sample preparation and, in many instances, analyses can be completed to the parts-per-billion level, which is not attainable with standard wet chemical techniques.

In some respects, atomic absorption does have limitations. In all cases, atomic absorption provides only the total concentration of the element. Unless specialized pretreatment is utilized, no breakdown of oxidation state or ionic species can be determined. However, for the analysis of metals at very low concentrations, atomic absorption is unsurpassed in speed and accuracy. In situations where a large number of metal analyses may be necessary, such as those POTWs that receive wastewaters from metal processing or finishing industries, atomic absorption capabilities may be required. But in all cases, the relatively high cost of atomic absorption equipment should be weighed against the need for high-volume trace-metal analysis.

Atomic emission spectrophotometry, commonly referred to as plasma emission spectrophotometry, is a relatively new analytical technique which can be used to analyze for as many as two dozen metallic pollutants simultaneously. Plasma units are more expensive and complex than standard atomic absorption, but for the municipality faced with a very high volume of metals analyses, these units may be applicable.

Specific-Ion Electrodes.   Specific-ion electrodes are sensing probes that can detect the concentration of chemical species when immersed in a solution containing the substance to be measured. As long as the probe is completely submerged, a concentration can be measured regardless of the volume of sample present. In contrast to atomic absorption, specific-ion electrodes, as the name implies, detect only certain ionic species as they exist in solution. As a consequence, specific-ion electrode readings are greatly dependent on the environmental conditions within the sample, such as pH and oxidation–reduction potential.

More than two dozen ionic species can be analyzed using specific-ion electrodes. Analyses are rapid, but require some pretreatment to remove interferences prior to immersion of the probe in the sample and meter readout. The drawbacks to this type of specialized equipment include possible fouling of the probe membrane, long readout equilibrium periods, and the ability to detect only specific ionic species. However, progress is continually being made in ion-sensing electrode technology. Excellent use has been made of the commercially available dissolved oxygen probes for measuring oxygen demand, and an ammonia-sensing electrode has been successfully used for monitoring nitrification in an activated sludge treatment plant. Because of the potential ease and speed of analysis that can be realized with specific-ion electrodes, consideration should be given to the possible use of these methods for selected ionic species.

Automatic Analyzers.   A few manufacturers currently market automated wet chemical analyzers that are typically called automatic analyzers. These devices automatically draw a small sample, add pretreatment chemicals, filter the sample if necessary, add chemicals to develop a color with an intensity proportional to the concentration, and finally automatically read and record the concentration proportional to the developed color. Other more sophisticated automatic analyzers may use infrared or ultraviolet spectrophotometric detection, and some use fluorometers or flame photometers. In effect, the typical automatic analyzer eliminates the steps that a technician would have to perform in carrying out typical wet chemical analysis.

Although an automatic analyzer can greatly reduce the time required to perform a particular analysis, use of these instruments is only economical when the analysis is required on a mass-production basis. Furthermore, automatic analysis instruments are relatively complex devices that can require a substantial maintenance effort. As indicated for the other specialized equipment, automatic analyzers should only be used when the presence of a specific pollutant contributed by a particular industrial source is so important that frequent analyses are required.

IR-UV Spectrophotometry.   Infrared (IR) and ultraviolet (UV) spectrophotometers, like gas chromatographs, are used to analyze for organic materials in wastewater. UV spectrophotometry has been used recently for detecting oil and grease in wastewater samples. However, these two types of specialized equipment generally have only limited and specialized uses because of cost and the requirement for skilled operators.

Gas Chromatography–Mass Spectrometry.   Gas chromatography refers to several variations on a technique in which a wastewater sample is concentrated and vaporized in a separation column packed with media having known affinities for particular organic molecules. The column is usuallly eluted with an inert gas and the times and intensities at which various organic materials exit the column are related to the specific organic compounds present in the sample. At times, gas chromatography cannot distinguish between similar organic materials. In these cases, gas chromatography must be coupled with mass spectrometry, in which the organic compounds are ionized, fragmented and sorted according to their masses. For a typical wastewater sample the evaluation of the raw data obtained from a GC–MS run may require interface with an on-site computer system. Gas chromatography–mass spectrometry is by far the most effective way of analyzing for trace organics in wastewater. However, most GC–MS systems are relatively expensive and require highly skilled operators. As a result, this type of equipment finds only limited application in POTW analytical laboratories.

Special Analytical Considerations

Underlying the compliance and enforcement uses of analytical data is the need to know the true composition of wastewater contributions so that the presence of harmful materials can be detected. In general, EPA recommends that wastewater samples be stabilized in a manner that will solubilize otherwise insoluble, or suspended materials. This is an especially important factor in the case of metals which are generally insoluble at high pH levels. Metal concentrations are of particular interest to plant operators, since relatively small quantities of these materials can cause operating problems. However, the metallic pollutant generally must be in solution before it can cause an upset of biological treatment processes. As a result, the practice of acidifying samples and obtaining total, rather than dissolved metal concentrations can give misleading results for samples with high pH levels. Metallic pollutants at high pH values would tend to be in the form of suspended solids, and would more likely be substantially removed prior to reaching biological treatment facilities. Consequently, the impact of a wastewater stream can greatly depend on the pH of the wastewater as it enters the treatment system.

Also important is the oxidation state of the constituent being measured. For example, hexavalent chromium is generally considered to have a greater impact on biological treatment processes than trivalent chromium. The impact of hexavalent chromium can be significant, but the typical environment in a sewer system would tend to reduce this material, and often ensure that only trivalent chromium reaches the treatment plant. Clearly, the analysis of wastewaters containing materials which depend on the conditions in the sample, must be approached prudently, considering the state of the pollutant as it enters the treatment system.

Personnel and Degree of Expertise

It can be stated in general terms that any industrial monitoring program will require an added degree of laboratory support. For very small systems with few industrial contributors, the additional analytical work can probably be handled by the technician performing regular sanitary chemistry. Larger systems may require the addition of personnel to handle the greater load created by the industrial monitoring program. If analyses in connection with the monitoring program require the use of any of the specialized analytical equipment outlined above, technicians or chemists with more than the usual level of expertise may be required. Although the utilization of specific-ion electrodes is relatively simple, proper use of atomic absorption, automatic analyzers, gas chromatographs, GC–MS, or IR–UV spectrophotometers does require a higher degree of expertise. Automatic analyzers and atomic absorption require good technique, and special attention to equipment maintenance. In order to properly utilize either GC–MS techniques or IR–UV spectrophotometry, special technical expertise beyond the four-year college level may be required.

Correlation of Analytical Techniques

In determining compatible pollutant loadings, it is necessary to analyze industrial wastewater contributions for the typical oxygen demand parameters. BOD historically has been used to measure oxygen demand, but this test is time consuming and can be inaccurate. COD has been used in place of BOD, but it too can be tedious, and it also takes a significant amount of time to complete. In recent years, catalytic oxidation approaches have been developed that yield measures of oxidizable substances that can be correlated to the more standard BOD or COD tests. These analyses, which are termed total organic carbon (TOC) and total oxygen demand (TOD), are rapid and reproducible.

A large body of data has been developed showing that the more rapid TOC and TOD methods can be correlated to BOD, and therefore can act as a more efficient measurement tool for an operating facility. Other similar correlations can be developed that can increase the efficiency of laboratory operations. For example, oil and grease is usually measured by an extraction gravimetric technique. This measurement can be simplified or accelerated by utilizing a UV spectrophotometer to detect the levels of organics if this type of device is available. Substitution of methods that are either faster, more accurate, or encompass a broader range of parameters should be considered for unofficial use, when they can be correlated with standard methods.

Standard Reporting Procedures

Laboratory data handling and reporting require two basic considerations; reliable methods for recording both laboratory and field data, and criteria for determining the significance and acceptability of the data. Without a good system in which standard procedures are used for accurate recording of analytical and field data, the usefulness of the information obtained from monitoring operations can be greatly diminished. Standard forms should be developed for recording field information, which would include the conditions at the time of sampling. All laboratory data should be recorded in bound notebooks with numbered pages. This assures a continuity in time, with a sequence for all analytical data. All forms should be completed in duplicate, with each copy being stored separately as a precaution against accidental loss of data.

Of equal importance to proper recording methods is the significance and acceptability of data collected. One must be certain that the sample being analyzed is representative, and has not been influenced by infrequent or rare laboratory or plant occurrences. Numerous statistical techniques are available that can provide a measure of the appropriate confidence that should be assigned to the data. These statistical techniques are adequately covered in several of the handbooks cited above. However, caution should be exercised when using statistical methods.

Generally, statistics offer a means by which variations in a set of data can be analyzed, assuming that all of the data used in the analysis are equally valid. Although statistics can be a powerful tool, it is always preferable to evaluate the significance of data variation on the basis of first-hand knowledge of the situation during sampling and analysis. Many times, if unusual circumstances are apparent, the data can be eliminated from consideration by inspection, and the problem can be rectified without the use of more sophisticated statistical analysis.

Contracting For Analytical Services

Many of the considerations discussed above concerning laboratory aspects of a monitoring program are based on the assumption that a POTW either has an existing laboratory which will be expanded to handle industrial wastewater, or has the resources to develop such facilities. This may be true for the larger systems, but the more numerous small systems may not have the resources or capability to complete the necessary analyses required for proper monitoring of industrial wastes. In such systems, analytical work must frequently be performed by commercial laboratories. Several criteria should be considered in choosing a laboratory to ensure that proper services are being provided.

At the present time only a few states have developed a certification system for commercial laboratories. EPA is currently in the process of developing a guidance document for lab certification programs. As a result, for the short term, POTWs wishing to engage qualified laboratory services will be required to evaluate laboratory performance independently. For the long term, use may be made of the upcoming guidance to be supplied by EPA.

Several techniques can be used to compare the quality of analytical services provided by commercial labs:

—*Use of samples spiked with known amounts of pollutants of interest.* Chemicals used in the spiked samples can be obtained from the appropriate EPA regional quality control coordinator. Use of spiked samples is a desirable method for testing laboratory performance when analyzing complex wastewater mixtures.

—*Parallel analysis of identical samples at two or more commercial labs.* This procedure can provide information on the relative performance of the laboratories in question.

—*Adherence to standard procedures.* This is essential, and as a result should be used as a primary criterion for evaluating lab performance.

—*Competitive costs.* These are also important, providing that an acceptable quality of analytical services is provided.

USER CHARGE SYSTEMS

Introduction

The total revenue requirements of a POTW generally fall into two broad categories:

1. Operation, maintenance, and replacement cost recovery;
2. Construction of capital cost recovery.

Traditionally, the funds necessary to cover these costs are raised using two methods: taxation based on assessed valuation (or other similar factor) and service charges related to the quantity and quality of contributions to the sewer. In recent years, increasing federal and state involvement in local POTW affairs has resulted in a complex set of rules and regulations governing various aspects of POTW revenue collections. Prominent in this area are federal and some state requirements that recipients of construction grant monies must insitute user charge and industrial cost recovery systems. Additionally, federal pretreatment regulations (1) state that any POTW required to set up a pretreatment program is also required to demonstrate adequate financial support for the program. The meaning of the term *adequate* is broad, and, in fact, financial support for the pretreatment program need not be linked to either a service charge or revenue based on assessed valuation. Instead, the financial arrangements must only be demonstrated to be capable of supporting the industrial waste control program. Within this context many revenue-producing techniques could be conceived to be adequate, but no particular means of financial support for an industrial control program are pinpointed by the federal pretreatment regulations.

It should be noted, however, that under the federal pretreatment regulations, federal funds have been made available to cover 75% of the costs of developing a POTW pretreatment program. Under sections 201 and 208 of the Clean Water Act, POTWs may apply for grant monies to develop industrial waste control programs. As a result, any POTW receiving federal support for the pretreatment program development will be required to abide by the user charge and industrial cost recovery regulations applicable to all grantees.

If the financial burden of the industrial waste control program is proportionately divided among the industrial contributors that the program is designed to manage, not only will general equity be achieved, but to some extent industrial contributions might be diminished and goodwill can be maintained. As a result, the method of choice for financing the industrial control program, keeping in mind the concept of equitable basis, is to incorporate the revenue generation needs of the industrial control program into the POTWs overall user charge system. Further, that user charge system should be structured to impact proportionately on all users and beneficiaries of POTW facilities and operations.

Definitions

Historically, many different approaches to POTW revenue production have been developed. As a result, various terms relating to the system of charges adopted by POTWs have come into usage. The terms generally associated with the subject of municipal user charges include *user charge, surcharge, sewer rental, industrial cost recovery, ad valorem tax,* and *special assessment.*

User Charge

User charges refer to a form of service charge that traditionally has many connotations, including the overall charges to an industry for the use of a POTW facility. However, in the context of PL 92-500 and the Clean Water Act of 1977, user charge assumes a specific denotation. User charges are defined as those charges assessed against contributors to the treatment works for their proportional share of operation and maintenance (including replacement) costs. The system must fulfill the objective of distributing the costs of operation and maintenance among all users in proportion to their waste load contributions. That is, those industrial contributors that discharge either high-strength or high-volume wastewaters would be required to pay a correspondingly higher proportion of total operation and maintenance costs. Specific factors such as BOD or flow for a particular contributor may be compared to overall system flow or BOD loading, and charges levied in accordance with the impact of the specific user as compared to the total system. Under current rules and regulations, the concept of user charges has been modified somewhat in that it is now acceptable for a federal grantee to fund operation and maintenance using property or ad valorem taxes. Nevertheless, the traditional meaning of user charges remains the proportionate recovery of operation, maintenance, and replacement costs.

Industrial Cost Recovery

Industrial cost recovery is another form of service charge. With industrial cost recovery, industrial users of a publicly owned treatment works repay the proportional federal share of the construction cost of the treatment works allocated to their use. Capital costs are to be recovered by the POTW, as a grantee of federal funding, over the useful life of the treatment works (not to exceed 30 years). The system is applicable only to the federal share of the construction cost. There is no federal requirement for recovery of the state or local share of costs for treatment of industrial wastes, although state or local laws may be so enacted. The federal share of the cost of construction includes the step 1, 2, and 3 grants, except for the costs associated with infiltration and inflow analysis, sewer rehabilitation, and nonexcessive infiltration and inflow if they are not attributable to industrial users.

Industrial cost recovery assessments are levied in proportion to the industrial user's wastewater characteristics. However, in the past several years municipalities have strongly opposed this assessment for small dischargers. Under the Clean Water Act of 1977, congress has agreed to exempt those industries with less than 25,000 gpd of wastewater, as long as the discharge does not include pollutants that would contaminate the plant's sludge. Further study of industrial cost recovery for larger contributors was also mandated by congress.

Ad Valorem Tax

The ad valorem tax is a charge for sewer use, both domestic and industrial, that is based typically on property value. As such, the cost of sewer service is generally combined with other municipal costs and financed from the municipality's general fund. The tax might also be based on the property area (square footage) or less equitable factors determined on a local basis.

Surcharge

The term surcharge is another traditional expression used to denote a service charge imposed on industrial contributors to a POTW based on quality and quantity of wastewater beyond the charge for normal domestic sewage. A typical arrangement found at many POTWs is one in which all contributors, domestic, institutional, or industrial are required to pay a fixed charge based on flow and the strength of a normal residential contributor in the system. That basis of wastewater strength generally would be developed as a result of the initial and ongoing industrial waste survey. The surcharge would then be an additional charge based on the increased strength or flow of an industrial wastewater as compared to typical residential discharge characteristics. If the rate structure is appropriately devised, surcharges can provide the added benefit of encouraging industries to reduce the volume and strength of their contributions.

Sewer Rental

Sewer rental is another loosely defined term commonly used for a service charge imposed on industry by a publicly owned sewage agency. It generally is computed on a volume-of-wastewater basis, and typically includes a standard hookup charge.

Special Assessment

The term special assessment is a kind of catchall method used in the past to develop sewer charges based on factors such as front footage or water consumption, and at times contract agreements between the POTW and individual sewer system users.

User Charge Strategies

Although there are various terminologies covering the subject of user charges, the means through which the pretreatment program cost may be equitably passed back to the industrial contributors fall into two categories: service charges and general revenue based on property value or other similar factors. Whatever basis is chosen, it will usually be part of the larger program used to finance the overall POTW operations. Since the cost of the pretreatment program has both operation and maintenance and capital cost elements, the choice of a user charge strategy should be linked to the overall financial strategy for the POTW.

In embarking upon a program to develop a revenue system to support the POTW, an evaluation of the relative merits of service charges or property taxes must be made. Perhaps the simplest means of financing the POTW operation is through property taxes. Traditionally, governmental activities are funded through property or ad valorem taxes. POTW operations can also be supported by drawing from the general fund. As expenses rise, the assessments on property may also rise keeping pace with increased POTW costs. Using ad valorem taxes to finance the POTW is favored by many contributors because, unlike service charges, taxes are deductible against income. The drawbacks to this type of financing include the fact that ad valorem taxes may be disproportionately levied against dry industries for support of the POTW operation. Likewise, heavy dischargers may end up contributing a lower than appropriate portion of the overall costs because of lower assessed value.

Service charges based on water use, volume of discharge, wastewater characteristics, and other factors, may offer a solution to some of the problems associated with property tax financing. By linking charges to quality and quantity of wastewater, the inequities associated with ad valorem taxes can be avoided. In this type of arrangement, each contributor would be charged on the basis of the relative benefit derived from discharge to the sewer regardless of tax exemptions (as in the case of institutions or churches). Service charges are not always popular because unlike taxes they generally are not deductible against income. Further, service charges cannot adequately finance the costs associated with excess capacity for future use installed to cover undeveloped properties or the added costs caused by infiltration. However, on the positive side, when an industry perceives a direct relationship between wastewater quality and quantity and the cost of discharge to the sewer, service charges can encourage voluntary abatement of discharges.

The key decision that emerges is whether to use ad valorem taxes, service charges, or a combination of the two. Because of statuatory limitations, particularly with regard to use of ad valorem taxes by federal grantees, not all POTWs are free to use any revenue system. Ideally, however, a combination of the two would provide the most effective type of system.

System Development

The first step in system development is to determine the actual cost of operating the POTW. A detailed accounting is usually required in which major areas of expenditure are delineated. These areas should include a breakdown of operation, maintenance, construction, and administrative costs for specific unit processes and specialized operations within the POTW. At the minimum, an actual dollar amount would be assigned to operation of primary, aeration, secondary, sludge-handling, pumping, monitoring, and laboratory facilities. A similar accounting would then be developed for anticipated construction costs or capital payments.

If the POTW already has a user charge system in force, it is likely that the detailed accounting may be a routine part of the local administrative process. In such cases, only estimates of the incremental additional expenses associated with the pretreatment program need be developed. Typically, additional operating costs would come from increased office and laboratory staff as well as legal and consulting engineering services required to develop and operate the pretreatment program. Added field monitoring activities and special equipment such as vans, automatic samplers, or portable pH meters would be included in the estimates.

If a federal grant is obtained to fund development of the POTW pretreatment program, an accounting of the initial costs of the program must be made. The national pretreatment strategy, which is included as an appendix to the federal pretreatment regulations (40 CFR, part 403), lists those preparatory steps in program development that are recommended and are presumably grant eligible. The steps indicated as appropriate for developing a pretreatment program, and for which a detailed accounting must be completed, are quoted directly below (1):

1. Developing an inventory of industrial and commercial wastes being introduced into the treatment works;
2. An evaluation of legal authority, including the adequacy of enabling legislation and the selection of mechanisms to be used for control and enforcement;
3. An evaluation of financial programs and revenue sources to ensure adequate funding to carry out the pretreatment program;
4. A determination of technical information (including specific requirements to prevent sludge discharges and specify violations of the discharge prohibitions . . .) necessary to develop an industrial waste ordinance or other means of enforcing pretreatment standards;

5. Design of a monitoring enforcement program;
6. A determination of pollutant removals in existing treatment works;
7. A determination of the treatment work's tolerance to pollutants which interfere with its operation or with sludge use or disposal;
8. Purchase of monitoring and laboratory equipment for use by the POTW;
9. Construction of facilities necessary to monitor industrial wastes; and
10. Any other works or activities approved by the Regional Administrator as necessary to obtain approval of the pretreatment program pursuant to 40 CFR Part 403.

Little information on the actual cost of setting up and operating a pretreatment program exists. However, as might be expected, the cost of an industrial pollutant control program is a function of both the total system flow and of the proportion of the industrial wastewater contribution. A 1975 survey of municipal and regional sewage systems (4) with dry-weather flows ranging from 15 to over 1,000 MGD was conducted. The survey showed that the annual direct cost of the industrial pollutant control program was generally on the order of $1,000 to $1,500 per MGD of dry-weather flow for systems in which some significant portion of the total wastewater flow came from industrial contributors. On this basis, the direct cost of a program covering industries for a system with an average flow of 50 MGD might be $50,000 to $75,000 per year (1975 dollars). Assuming an average salary level of $15,000 per year for personnel assigned to such a program, this system would support a staff of 3–5 persons.

Following the detailed accounting, the costs should be allocated to the users or beneficiaries of the system. (Undeveloped properties may not be users but are beneficiaries of the system and should be required to bear an appropriate portion of the costs). Allocation of costs can be achieved by relating the design capacity of the POTW to actual pollutant load.

A major result of the initial and ongoing monitoring program should be a rough mass balance for the design parameters within the collection system. Using this information, pounds of pollutant contributed to the sewer can be calculated. Knowing the pounds of pollutant contributed or the volume of water discharged provides an opportunity to link these parameters to the unit operations designed to handle them.

The overall cost of a particular unit process or operation has both an operation/maintenance and a capital cost element. Additionally, these two cost elements may be divided into current and future use segments. The operation, maintenance, and construction of a pumping station can provide a simple example of the application of design parameter load to revenue production. Using the total volume of wastewater handled by the pumping facility and its overall design capacity, the costs of the system can be proportionately allocated. Since the pumping station may not be working at full design capacity, the costs may be split into current use and future use sectors. For current use, the proportion of the total flow pumped as related to the actual gallons contributed by an individual discharger can be used to determine the contributor's service charge to support the pumping station. The excess capacity in the pumping facility represents future use which is not clearly attributable to any particular user or beneficiary. Consequently, a blanket charge such as property taxes or ad valorem taxes would be most appropriate to cover the maintenance and capital or construction cost of the excess capacity.

Similarly, for unit processes within the treatment plant where there are multiple design parameters, operation, maintenance and construction costs can be recovered proportionately. If, for example, aeration basins are designed to remove BOD, the cost of operating, maintaining, and constructing these facilities should be linked to pounds of BOD treated daily. However, for the case where the operation unit process is designed on the basis of more than one design parameter, such as BOD and suspended solids, a split between the parameters should be made for allocating the costs. The proportion of this split may not be clear and as a consequence a subjective judgment may be required.

The end product of these calculations is usually a set of formulas used to calculate service charges. Additionally, costs associated with nonattributable items such as future use and infiltration would be worked into the property tax rate. Common practice is to derive an overall formula containing several major parts. The basic factor in the formula is a service charge paid by all dischargers to the sewer. This base charge, which is often linked to volume of discharge, should be structured in such a way that it would cover the cost of handling all contributions if they were all no stronger than standard domestic sewage. By instituting a domestic user charge of this type, routine evaluation of residential sources is not needed once the concentration and loading of standard domestic wastewater are established.

For industries that have waste loads greater than domestic waste, the formula should contain a surcharge factor. The surcharge formula is designed to charge incrementally for each pound of pollutant above the standard domestic waste load. If an industry has a weak wastewater at or below domestic strength, the surcharge formula is designed to yield no additional charge.

Finally, to cover the costs of excess capacity, infiltration, and other nonattributable items, the formula should contain a property tax element. Normally a tax rate is calculated which is multiplied by the assessed value to determine the contribution to the POTW.

In summary, utilizing the concepts of proportional payment and combined service charges and property value taxes, all members of the community, whether users or simply beneficiaries of the POTW services, will contribute some funds to the POTW operation. Because the pretreatment program costs are a result of industrial contributions, they should be equitably charged back to the industrial sector. However, the total costs of the POTW operation should be shared fairly by all property owners.

Case History–Buffalo Sewer Authority (2)

The problems involved in developing a method by which the Buffalo Sewer Authority (BSA) was to distribute its financial obligation to its beneficiaries were quite complex, and perhaps comparable to the problems of other municipalities of similar size and age. The guiding principle in the development of a method by BSA was that the total annual revenue required for the operation of the sewage works should be distributed to the users of the system. In addition, the distribution should be approximately in proportion to the cost of providing the use and benefits to each user. The philosophy used in Buffalo was that the cost should be distributed according to the benefit and use made of the system. For individual users, such factors as community benefit, readiness to serve, and ability to pay were not considered.

The cost distribution method approved by EPA for the Buffalo system included a volume and loading base for the costs resulting from point source discharges, both domestic and industrial, and an ad valorem base for the non–point source discharges in the form of infiltration and future use. It should be noted that while EPA approved this method under PL84-660 regulations, and would disapprove it under PL92-500, under the Clean Water Act of 1977 EPA would now approve of the ad valorem tax base.

Distribution of Costs by Design Parameters

The BSA's secondary treatment plant has a number of design considerations (flow, BOD, SS, PO_4^{3-}) and the relative proportions charged to users and property owners differ for the various unit processes. For example, the pumping station, final settling tank, chlorine tank, plant outfall, and modification to the gate chamber were all designed based on flow. Consequently, the costs were distributed on the basis of flow. On the other hand, the grit disposal facility, the modification of the primary tanks and the primary digestors were designed based on suspended solids loading. Since the capital costs as well as the operation and maintenance costs of the treatment facility are directly related to the parameters for which it was designed, the proportional allocation of the cost of the works to the users of the system or the property owners was based on these same parameters.

To derive the formula for the allocation of the estimated first year operation and maintenance (O & M) costs, a new plant was broken down into 15 unit operations (Table 3–3) with each being assigned its portion of the total O & M cost relative to the appropriate design parameters. The apportionment resulted in: 25.5% of the total O & M cost assigned to flow; 41% to suspended solids removal; 17.5% to BOD removal; and 16% to phosphate removal.

The allocation of capital construction costs was arrived at in much the same manner (Table 3–4) with the costs of twenty individual facilities each being assigned its portion of the total cost relative to the appropriate design parameters. This breakdown resulted in 30.8% of the total capital cost being attributed to flow; 25.3% to suspended solids removal; 41.6% to BOD removal; and 2.3% to phosphate removal.

Based on estimated material balances for the design parameters in the Buffalo system and on the plant's design criteria, the loadings received at the plant were distributed to future use, infiltration, residential use, and industrial use. According to this distribution then: 27% of the first year capital cost would be attributed to future use, 9% to infiltration, 46% to residential users, and 18% to industrial users, as shown in Table 3–5.

The estimated first-year O & M costs were distributed in a similar fashion, with the notable exception that none of these costs were attributable to future use. Therefore, 12% of the O & M costs were assigned to infiltration, 59% to residential users, and 29% to industral users, as shown in Table 3–6.

Combining the distribution schemes for the capital construction cost and the O&M cost, a total of 19% was chargeable to property based on future use and infiltration, and 81% was chargeable to the industrial and residential users, as shown in Figure 3–8.

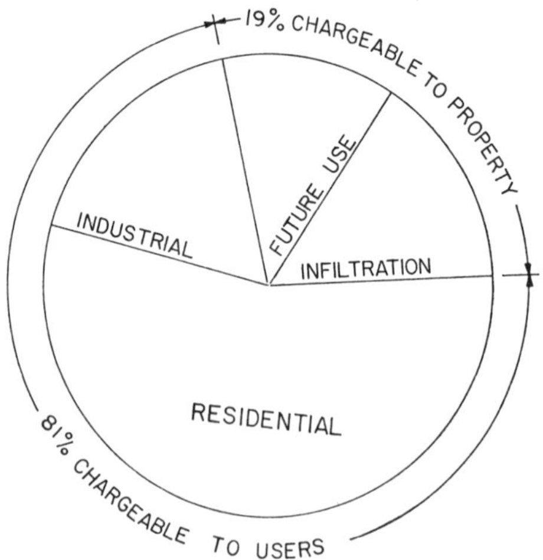

Figure 3-8 / BAS system: First-year cost distribution capital and O&M (2).

Table 3-3 / Distribution of BSA's O & M costs by design parameters (2)

Facilities	Breakdown By Design Parameters				
	Flow	SS	BOD	PO_4^{3-}	Total
Raw wastewater pump station	$ 267,300	$ —	$ —	$ —	$ 267,300
Administration building	162,100	139,000	139,000	$ 23,100	463,200
Grit tank and building	—	137,900	34,500	—	172,400
Sedimentation tank and building	—	98,500	98,500	—	197,000
Settled wastewater pump station	409,700	—	—	4,100	413,800
Thickener and filter building	—	1,078,500	—	—	1,078,500
Incinerator building	—	3,255,700	—	—	3,255,700
Blower building	—	—	985,400	—	985,400
Chemical building	1,905,900	79,400	—	1,985,300	3,970,600
Aeration tank and building	—	—	301,700	—	301,700
Final tank and building	—	92,400	92,400	—	184,800
Chlorine and traveling screen building	235,600	54,300	72,500	—	362,400
Digester control building	16,400	131,500	16,400	—	164,300
Laboratory	—	93,800	147,400	26,800	268,000
Miscellaneous	573,400	573,400	573,400	191,200	1,911,400
Total	$3,570,400	$5,734,400	$2,461,200	$2,230,500	$13,996,500
Percentage of Total	25.5%	41%	17.5%	16%	100%

Table 3-4 / Distribution of BSA's construction costs by design parameters (2).

Facilities	Breakdown by design parameters				
	Flow	SS	BOD	PO_4^{3-}	Total
Modifications to primary	$ —	$ 2,410,430	$ —	$ —	$ 2,410,430
Grit disposal facilities	—	170,660	—	—	170,660
Settled wastewater pump station	7,698,680	—	—	—	7,698,680
Secondary treatment units	—	—	34,066,760	—	34,066.760
Blower building	—	—	7,841,010	—	7,841,010
Final settling tanks	18,587,080	—	—	—	18,587,080
Chlorine contact tanks	5,750,810	—	—	—	5,750,810
Sludge concentration tanks	—	6,836,410	721,620	—	7,558,030
Sludge disposal facilities	—	17,610,860	1,858,920	1,808,670	21,278,450
Plant outfall sewer	656,600	—	—	—	656,600
Gate chamber	331,440	—	—	—	331,440
Phosphate removal facilities	—	—	—	647,920	647,920
Preliminary site preparation	3,875,580	3,183,520	5,234,550	289,410	12,583,060
Roadways and grading	1,109,790	911,610	1,498,940	82,870	3,603,210
New electric service	445,450	365,900	601,650	33,260	1,446,260
Heating and ventilation	1,299,220	1,067,220	1,754,790	97,020	4,218,250
Administration building	574,440	471,860	775,870	42,900	1,865,070
Miscellaneous structures	1,612,970	1,324,940	2,178,550	120,450	5,236,910
Maintenance shop	459,370	377,340	620,450	34,300	1,491,460
Technical and legal costs	7,255,840	5,960,150	9,800,090	541,830	23,557,910
Total	$49,657,270	$40,690,900	$66,953,200	$3,698,630	$161,000,000
Percentage of total	30.8%	25.3%	41.6%	2.3%	100%

**Table 3-5 / BSA system: First-year
capital cost distribution (2)**

| | Percentage Chargeable to: | | | |
| | Property | | User | |
Parameter	Future Use	Infiltration	Residential	Industrial
BOD	28	4	43	25
SS	50	3	34	13
PO_4	35	13	26	26
Flow	6	21	60	13
Total	27	9	46	18

**Table 3-6 / BSA system: First-year
O&M cost distribution (2)**

| | Percentage Chargeable to: | | | |
| | Property | | User | |
Parameter	Future use	Infiltration	Residential	Industrial
BOD	0	6	60	34
SS	0	5	68	27
PO_4	0	21	40	39
Flow	0	23	63	14
Total	0	12	59	29

Domestic Rate Formula

Having established the allocation of capital and O&M costs based on design
parameters, and in turn the characteristics of the individual waste actually
being discharged, the groundwork was laid for the determination of unit
rates, rate formulas, and charges. However, it was neither administratively
nor economically feasible to measure the strength of discharge and charge
each individual user of the system independently. Therefore the BSA established as a standard for domestic-strength waste 250 mg/l BOD, 250 mg/l SS,
and 15.35 mg/l PO_4^{3-} to be used in establishing the basic sewer use
charge. In addition, since high-strength wastes are directly responsible for
increased costs at the treatment plant, an extended charge or surcharge over
and above the basic sewer use charge was levied against industries discharging wastes with strengths above the domestic waste standard.

Based on the allocation of loading previously described, rate formulas
were developed. There are two basic formulas, the first for computing the

charge to the customers who discharge only domestic strength waste and the second for computing the rates charged to the industrial user. The domestic rate formula consists of two terms and is stated as follows:

Domestic treatment charge = Domestic user charge + Property charge.

Each of the above terms contains a portion of the capital costs as well as of the operating and maintenance costs.

By using the domestic-strength waste characteristics, all the unit rates for user charges could be expressed in terms of flow as follows:

Domestic user charge = $0.078/1000 liters
Industrial user charge = $0.106/1000 liters.

The industrial user charge was set higher than the domestic charge to account for the additional administrative and engineering costs associated with the established industrial waste control program.

Using the appropriate unit cost and the total assessed valuation of taxable property in the city of Buffalo, a property rate of $2.32/$1000 and $1.75/$1000 of assessed valuation for the capital and operations portions, respectively, was derived. The formula equates charges in dollars to water consumption in thousands of liters.

Industrial Rate Formula

The industrial rate formula is an expanded version of the formula for domestic discharges. It consists of three basic terms, as follows:

Industrial waste treatment charge = Domestic user charge
+ Property Charge + Surcharge.

The simple addition of the surcharge term makes the formula broadly applicable to all industries. When applying it to an industry that discharges only domestic-strength waste, the last term drops out and the formula becomes that of a domestic user. The surcharge portion, as stated below, provides an additional charge for each milligram per liter of BOD, SS, and PO_4^{3-} being discharged over and above what is considered domestic-strength waste:

Surcharge = $Q(0.152(\text{BOD} - 250) + 0.291(\text{SS} - 250) + 1.63(PO_4^{3-} - 15.35)$,

where Q is the average annual discharge rate expressed in million liters/year.

The determination of waste characteristics that would define an upper limit for domestic-strength waste required a great deal of investigation. Ordinances from other municipalities across the country were reviewed and it was found that the range of values adopted varied from 250 mg/l to 350 mg/l for BOD and suspended solids. Since the Buffalo metropolitan area has always had what would be considered a weak waste, it was deemed appropriate to use the low values.

In conclusion, if the industrial user charge and property charge are substituted in the above formula, the final form is as shown below, where Q_I is the industrial wastewater flow in thousands of liters per year, and A is the assessed property valuation in thousands of dollars:

$$\text{Yearly industrial waste treatment charge (\$)} = Q_I(0.106) + A(4.071) + \text{Surcharge}$$

COMPLIANCE PROCEDURES

The POTW control agency must develop procedures to ensure compliance with the POTW pretreatment program. These procedures should at the minimum provide for (1):

—Systematic, comprehensive surveys of collection systems to identify and locate all appropriate IU's; this inventory shall be made available to the state upon request;
—Identification of pollutant discharges by character and volume by the IUs identified by the survey;
—Notification of IUs identified by the survey of applicable pretreatment standards and sludge management requirements;
—Receiving and analyzing the self-monitoring reports and other notices submitted by IUs;
—Sampling and analyzing at random the effluents from IUs;
—Conducting surveillance and inspection activities to ensure compliance;
—Following up conditions of noncompliance and collecting additional data, samples, etc., sufficient to provide admissible evidence for enforcement proceedings;
—Initiating court actions for injunctive relief, etc., for noncompliance;
—Implementing and enforcing other applicable rules, regulations, etc.;
—Complying with requirements for public participation, notification, etc.; this includes public notice at least once every twelve months in the largest local daily newspaper, of all IUs who were not in compliance with pretreatment standards or other applicable pretreatment requirements during the preceding year; a summary of enforcement actions during the same twelve-month period shall be included in the notification.

Contents of the Compliance Schedule

The compliance schedule should require that the permittee develop the authorities, procedures and resources which comprise an approvable POTW pretreatment program. The activities listed in the model compliance schedule (Fig. 3–9) summarize the more detailed requirements of the pretreatment regulation. It is recommended that the state review the more detailed requirements set forth in the regulations before developing the pretreatment compliance schedule, and insert additional schedule activities where appropriate.

The terms and conditions of the POTW pretreatment program, when approved, shall be enforceable automatically through the permittee's NPDES permit. Activity 1 of the model compliance schedule (requiring the POTW to apply for a 201 grant amendment to acquire pretreatment funding) must be included whenever the POTW's permit is modified to include the pretreatment compliance schedule before the permittee has acquired funding to carry out the program. Construction grant funding cannot be used to pay for work done by the POTW before receipt of the 201 grant amendment. The state (approval agency) may consider the option of inserting in the municipal permit, along with the compliance schedule for the development of a pretreatment program, a clause stating that if the industrial waste inventory required by the compliance schedule (activity 2) demonstrates that the POTW has no contribution of industrial wastes which would be subject to pretreatment requirements, the POTW would not be required to continue development of the program.

Developing Compliance Schedule Dates

There are several time limitations imposed by the pretreatment regulation and the construction grant regulation (40 CFR, Part 35) which should be considered in establishing compliance schedule dates. The pretreatment regulation provides that the compliance schedule will require the development and approval of a POTW pretreatment program as soon as is reasonable and within three years after the schedule is incorporated into a POTW's permit, but in no case later than July 1, 1983. Since up to six months must be allowed for the program approval process, the compliance schedule date for submission of the pretreatment program for approval (activity 9 of the compliance schedule) should be two and one-half years from the incorporation of a compliance schedule or January 1, 1983, whichever is sooner.

Provisions of the construction grants regulations impose what may be in some cases stricter time constraints on the development of an approvable program. The construction grants regulation provides that no guarantee may receive a step 3 grant after December 31, 1980, until it has developed an approvable pretreatment program. Thus, a permittee which is scheduled to receive a step 3 construction grant in January 1981 will be required to develop an approvable program at the outside by January 1981. However, if that same permittee received a compliance schedule for the development of a pretreatment program in December 1978, it would be allowed by the pre-

ACTIVITY NUMBER	ACTIVITY	COMPLETION DATE
1.	The permittee shall apply for an amendment to its existing step 1, 2, or 3 construction grant to acquire funding for the development of a pretreatment program.	________
2.	Submit the results of an industrial user survey including identification of IUs and the character and volume of pollutants contributed to the POTW by the industrial users.	________
3.	Submit an evaluation of the legal authorities to be used by the permittee to apply and enforce the requirements of the Clean Water Act, including those requirements of the pretreatment regulation.	________
4.	Submit a determination of technical information necessary to develop an industrial waste ordinance or other means of enforcing pretreatment standards.	________
5.	Submit an evaluation of the financial programs and revenue sources which will be employed to implement the pretreatment program.	________
6.	Submit design of a monitoring program which will implement the requirements of the pretreatment regulation.	________
7.	Submit a list of monitoring equipment required by the POTW to implement the pretreatment program and a description of municipal facilities to be constructed for monitoring or analysis of industrial wastes.	________
8.	Submit specific limitations for prohibited pollutants contributed to the POTW by IUs.	________
9.	Submit a request for pretreatment program approval and removal credit approval, if desired.	________

Figure 3-9 / Model compliance schedules.

treatment regulation an outside date of June 1981 (i.e., two and one-half years from the incorporation of the compliance schedule) to develop an approvable program. In this case, the more stringent time limitation (i.e., that posed by the construction grant regulation) would apply.

As the example above indicates, in developing the schedule date for the submission of an approvable pretreatment program, the state must use that date prescribed by either the pretreatment regulation or the construction grants regulation which provides the shortest time for the development of the program. In addition, the State may impose reasonable time limitations which are more restrictive.

Regulatory limitations on the time frame for developing a program can be summarized as follows:

1. Approval within three years from the incorporation of a pretreatment compliance schedule in the municipal permit (application for approval within two and one-half years);
2. Approval by July 1, 1983 (application for approval by January 1, 1983);
3. Approval prior to payment of grants beyond 90% of the step 3 funding (application for approval six months before this date);
4. Development of an approvable pretreatment program by the end of the step 2 grant for certain permittees;
5. Approval by whatever more stringent time limit is imposed by the permit-issuance authority.

In addition, the construction grant regulation imposes an interim time limitation on the development of compliance schedule activities 2–4. According to this regulation, grantees with amended step 1 grants must have completed activities 2–4 by the time of application for step 2 grant if the step 2 is to be awarded after June 30, 1980.

Facilities required to develop a POTW pretreatment program can generally be divided into four groups depending upon the applicability of the time limitations discussed above. (See Table 3–7).

Group 1: facilities which will have received step 1 and 2 construction grants or amendments before June 30, 1980, and a step 3 construction grant before December 31, 1980.

If a grantee is scheduled to receive its step 2 and 3 construction grants before June 30, 1980 and December 31, 1980, respectively, the construction grant regulation requires that, in most cases, the grantee have an approved POTW pretreatment program before it receives the last 10% of its step 3 grant funding. This means that the grantee would be required to apply for POTW pretreatment program approval at least six months before it is scheduled to receive payment beyond 90% of its step 3 funding. The pretreatment regulation provides that such a grantee should request approval of the POTW pretreatment program within two and one-half years from the incor-

Table 3-7 / Outside pretreatment compliance dates based on construction grant awards and pretreatment requirements.[a] (1)

Group	June 30, 1980			December 31, 1980			2½ years from initiation of compliance schedule, Jan. 31, 1983, or 6 months before the final 10% of step 3 grant whichever is sooner
1 *Step 1* Awarded	*Step 2* Awarded		*Step 3* Awarded				Activities 2–9 Due
2 *Step 1* Awarded	*Step 2* Awarded				*Step 3* Awarded	Activities 2–8 due by application for step 3	Activity 9 Due
3 *Step 1* Awarded		*Step 2* Awarded	Activities 2–4 due by application for step 2	*Step 3* Awarded			Activities 5–9 Due
4 *Step 1* Awarded		*Step 2* Awarded	Activities 2–4 due by application for step 2		*Step 3* Awarded	Activities 5–8 due by application for step 3	Activity 9 Due

[a]Interim dates are negotiable and are established by the permit-issuance authority.

poration of a pretreatment compliance schedule into its NPDES permit, or by January 1, 1983, whichever is sooner.

In developing the compliance schedule for permittees in this group, states should determine which of the above dates provides for the earliest development of a POTW pretreatment program. This date should then be used as the pretreatment compliance schedule deadline for activity 9. Dates for the remaining compliance schedule activities are negotiable with the permittee. Generally, however, the deadlines for completing activities 2–4 should not exceed 15 months from the initiation of the compliance schedule. Facilities receiving their step 3 grant before June 30, 1980, shall be subject to the same time limitations described above.

Group 2: facilities which will have received step 1 and 2 construction grants before June 30, 1980, and a Step 3 construction grant after December 31, 1980.

The construction grant regulation provides that a grantee which is scheduled to receive a step 3 grant after December 31, 1980, must have completed compliance schedule activities 2–8 before it can receive its step 3 funding. Therefore, in developing the compliance schedule, the state should use as an outside compliance date for activities 2–8 the date for completion of the step 2 grant as determined by the construction grants compliance schedule, as long as this date would not be later than two and one-half years from the initiation of the pretreatment compliance schedule or January 1, 1983, whichever is sooner.

The compliance date for pretreatment compliance schedule activity 9 (request for program approval) should not exceed two and one-half years from the initiation of the compliance schedule, January 1, 1983, or six months before the permittee is scheduled to receive payment beyond 90% of its step 3 funding, whichever is sooner. Again, the interim pretreatment compliance schedule dates are negotiable. It is recommended that the completion date for activities 2–4 not exceed 15 months from the initiation of the compliance schedule.

Group 3: facilities which will receive a step 2 construction grant after June 30, 1980, and a step 3 construction grant before December 31, 1980.

Under the construction grant regulation, in order to receive a step 2 grant after June 30, 1980, a grantee must first have completed activities 2–4 of the pretreatment compliance schedule. The state should therefore ensure that the compliance schedule dates for the completion of activities 2–4 do not exceed the scheduled date for the completion of the step 1 grant activities. The state may at its discretion impose a more stringent time limitation for the completion of these activities. It is recommended that the completion date for activities 2–4 not exceed 15 months from the initiation of the compliance schedule.

The construction grant regulation provides that grantees which will receive a step 3 grant before December 31, 1980, should have an approved pretreatment program in order to receive the final 10% of the step 3 grant funds. The final compliance date for activity 9 of the pretreatment compliance schedule therefore should be no later than six months before the date upon which the grantee is scheduled to receive payment beyond 90% of the step 3 grant funding, unless this date exceeds two and one-half years from the initiation of the compliance schedule, or January 1, 1983, in which case the final date for activity 9 should be no later than January 1, 1983, or two and one-half years from the initiation of the compliance schedule, whichever is sooner. The interim dates for activities 5–8 are negotiable with the permittee.

Group 4: facilities which will receive a step 2 construction grant after June 30, 1980, and a step 3 construction grant after December 31, 1980.

The construction grant regulation provides that in order to receive a step 2 grant after June 30, 1980, a grantee must first have completed activities 2–4 of the pretreatment compliance schedule. The state should therefore ensure that the compliance schedule dates for the completion of activities 2–4 do not exceed the schedule date for the step 2 grant application. The state may impose a more stringent time limitation for the completion of these activities. It is recommended that the completion date for activities 2–4 not exceed 15 months from the initiation of the compliance schedule.

In order to receive a step 3 grant after December 31, 1980, a facility in this category must also have completed compliance schedule activities 5–8. The final compliance dates for activities 5–8 should therefore be no later than the completion date for the facility's step 2 grant as determined by the construction grants schedule. If the scheduled completion date for the Step 2 construction grant activities is later than two and one-half years from the initiation of the compliance schedule or January 1, 1983, then the final compliance date for activities 5–8 should not exceed January 1, 1983, or two and one-half years from the initiation of the compliance schedule, whichever is sooner.

In establishing the pretreatment compliance schedule dates for activities 5–8, sufficient time must be allowed for the grantee to accomplish activity 9 (application for program approval) by January 1, 1983, two and one-half years from the initiation of the pretreatment compliance schedule, or six months before the permittee is scheduled to receive payment beyond 90% of its step 3 funding, whichever is sooner.

Table 3-8 / State options

State Program 1	State Program 2	State Program 3
Major POTWs have programs.	Majority of POTWs have programs.	Few, if any POTWs have programs.
State assumes responsibility for a moderate number of IUs.	State assumes responsibility for a minimum number of IUs.	State assumes responsibility for a maximum number of IUs.
State coordination effort required—moderate resources.	State overview effort required—few resources.	Detailed state involvement required—maximum resource investment.

NPDES STATE OPTIONS

Three options are available to an NPDES state in meeting the requirement of implementing a state pretreatment program. Table 3-8 shows the available options. The option that appears to be reasonable and effective in most cases is option 1, where the state utilized applicable POTWs to conduct local pretreatment programs and concurrently conducts a program for the balance of IUs not covered by POTW pretreatment programs. This option requires a coordination effort and maximizes use of existing personnel within municipalities with in-house knowledge of local conditions.

The following deadlines applied to NPDES state programs:

—March 27, 1979. State submission to EPA for pretreatment programs (applicable to states with existing statutory authority);
—March 27, 1980. State submission to EPA for pretreatment programs (applicable to states requiring legislative changes).

The state will be required to develop, in a format suggested by EPA, a description of the state pretreatment program for implementing and maintaining the requirements of 40 CFR, part 403. The program format will include a description of program procedures, funding requirements and the attorney general's statement setting forth legal authority.

REFERENCES

1. General pretreatment regulations for existing and new sources of pollution. *Federal Register,* vol. 43, no. 123, pp. 27736–73, June 26, 1978.
2. *Industrial Waste and Pretreatment in the Buffalo Municipal System.* U.S. Environmental Protection Agency Publication no. EPA-600/2-77-018, MCD 31, January 1977.

3. Williams, R. T. and R. J. Dolan. How to manage industrial inflow. *Water and Sewage Works,* pp. 46–49, December 1974.
4. *Federal Guidelines—State and Local Pretreatment Programs.* USEPA Publication no. EPA-430/9-76-017a, MCD-43, January 1977.
5. Eason, John E., Jay G. Kremer, and Franklin D. Dryden. Industrial waste control in Los Angeles County. *Journal of the Water Pollution Control Federation,* pp. 672–77, April 1978.
6. *Regulation of Sewer Use.* WPCF Manual of Practice No. 3. Washington, D.C.: Water Pollution Control Federation, 1975.
7. *Handbook for Monitoring Industrial Wastewater.* USEPA Technology Transfer, August 1973.
8. Shelley, Philip E. *Sampling of Wastewater.* Washington, D.C.: USEPA, Technology Transfer, June 1974.
9. *Wastewater Sampling Methodologies and Flow Measurement Techniques.* Harris, Daniel J. and W. J. Keffer. USEPA Publication no. 907/9-74-005, June 1974.
10. Guidelines establishing test procedures for the analysis of pollutants—Proposed regulations. *Federal Register,* vol. 44, no. 233, pp. 69464-575, December 1, 1976.
11. *Manual of Methods for Chemical Analysis of Water and Wastes.* USEPA, Technology Transfer, no. 625/6-74-003, June 1974.
12. *Standard Methods for the Examination of Water and Wastewater,* 14th ed. American Public Health Association, American Water Works Association and Water Pollution Control Federation, 1975.
13. *Annual Book of Standards, Part 31, Water, Atmospheric Analysis.* American Society for Testing and Materials, 1975.
14. *Sampling and Analysis Procedures for Screening of Industrial Effluents for Priority Pollutants.* USEPA, Environmental Monitoring and Support Laboratory, April 1977.
15. *Analytical Methods for the Verification Phase of the BAT Review.* USEPA, Effluent Guidelines Division, Office of Water and Hazardous Materials, June 1977.
16. *Handbook for Analytical Quality Control in Water and Wastewater Laboratories.* USEPA, Technology Transfer, June 1972.

GENERAL REFERENCES

ADMINISTRATION OF A PRETREATMENT PROGRAM

Black, H. H. Planning industrial waste treatment. *Journal of the Water Pollution Control Federation,* vol. 41, no. 7, p. 1277, July 1969.
Byrd, J. F. Combined treatment. *Proceedings of the 16th Industrial Waste Conference,* Purdue University, 1961, p. 92.
Byrd, J. F. Combined treatment—A coast to coast coverage. *Journal of the Water Pollution Control Federation,* vol. 39, no. 4, p. 601, April 1967.

Byrd, J. F. All parties can benefit from joint municipal–industry treatment. *Water and Sewage Works,* vol. 116, no. 11, p. IW 14, November 1969.

Enforcement Management System Users Guide. USEPA, NTIS no. PB 210-716, 210 pp., September 1972.

Gibbs, C. V. and R. H., Bothel. Potential of large metropolitan sewers for disposal of industrial wastes. *Journal of the Water Pollution Control Federation,* vol. 37, no. 10, p. 1417, October 1965.

Hall, R. M., Jr. An attorney reviews EPA's new pretreatment program. *Industrial Water Engineering,* pp. 8–14, September 1978.

Hickman, P. T. The joint municipal and industrial wastewater treatment approach—A case history. Paper presented at the Water Pollution Control Federation Meeting, Denver, Colorado, October 9, 1974.

Jackson, C. J. Management of industrial effluent disposal in Britain. *Journal of the Water Pollution Control Federation,* vol. 41, no. 12, p. 2020, December 1969.

Meriwether, G. B., Treatment of mixed industrial wastes at Bayport's industrial complex. *Journal of the Water Pollution Control Federation,* vol. 41, no. 3, p. 440, March 1969.

Munson, E. D. New concepts in industrial sewage collection. *Journal of the Water Pollution Control Federation,* vol. 36, no. 9, p. 1146, September 1964.

Ott, R., et al. Effects of the 1972 Water Pollution Control Act amendments on industrial waste monitoring in Anondaga County. Paper presented at the New York Water Pollution Control Association, January 1974.

Reed, P. E. Pollution abatement thru government-corporate cooperation. *Water and Sewage Works,* vol. 121, no. 9, p. 104, September 1974.

Reiter, W. M. Evaluation factors for joint waste treatment. *Pollution Engineering,* vol. 6, no. 12, p. 38, December 1974.

Rocheleau, R. F., and E. F. Taylor. An industry approach to pollution abatement. *Journal of the Water Pollution Control Federation,* vol. 36, no. 10, p. 1185, October 1964.

Swets, D. H., et al. Combined treatment at Kalamazoo—Cooperation in action. *Journal of the Water Pollution Control Federation,* vol. 39, no.2, p. 204, February 1967.

Traquair, W. C. Rx for industry: Regionalism. *Water and Wastes Engineering,* May 1973.

Wastes may not be a treat for pretreatment. *Chemical Week,* October 9, 1974.

Watson, K. S. The advantages of industrial–municipal wastewater treatment. *Journal of the Water Pollution Control Federation,* vol. 42, no. 2, p. 209, February 1970.

Waytenick, R. J. Cooperation helps Erie. *Water and Wastes Engineering,* p. 76, September 1973.

Webber, P. J. and R. C. Kausch. Delaware system moves ahead. *Water and Wastes Engineering,* p. 44, January 1972.

Wisely, W. H. The foundation of successful industrial waste disposal to municipal sewage works. *Proceedings of the 5th Industrial Waste Conference,* Purdue University, p. 360, 1949.

DATA BASE DEVELOPMENT

Greenberg, M. R. and R. Zimmerman. Estimating industrial water pollution in small regions. *Journal of the Water Pollution Control Federation,* vol. 45, no. 3, p. 462, March 1973.

Meers, J. E., et al. Development of industrial waste study for a municipality. *Journal of the Water Pollution Control Federation,* vol. 36, no. 12, p. 1501, December 1964.

Westfield, J. D., et al. Planning and execution of industrial waste treatability studies. *Proceedings of the 26th Industrial Waste Conference,* Purdue University, p. 832, 1971.

Williams, R. T. Classifying industrial wastewater emissions. *Water and Sewage Works,* vol. 121, no. 7, p. 86, July 1974.

LEGAL ASPECTS

Anderson, N. E. and B. Sosewitz. Chicago industrial waste surcharge ordinance. *Journal of the Water Pollution Control Federation,* vol. 43, no. 8, p. 1591, August 1971.

Buffalo Sewer Authority, Buffalo, New York. Sewer regulations of the Buffalo Sewer Authority.

Byrd, J. F. Municipal waste ordinances—The views of industry. *Journal of the Water Pollution Control Federation,* vol. 37, no. 12, p. 1635, December 1965.

California Water Pollution Control Association, Berkeley, California. Model wastewater discharge permit application questionnaire, October 1974. Model wastewater discharge ordinance, April 1974.

Calver, R. and T. Saxon. The sewer ordinance basics. *Water and Sewage Works,* vol. 121, no. 8, p. 54, August 1974.

Carpenter, C. B. Control of industrial wastes entering municipal sewers. *Proceedings of the 11th Industrial Waste Conference,* Purdue University, p. 1, 1956.

City of Akron, Ohio. Ordinance no. 499: Industrial wastes, regulations for non-acceptable, 1963.

City of Atlanta, Georgia. Sewer service charges and industrial waste surcharges, 1971. Standards of acceptability of industrial or trade wastes for admission into sewers of the City of Atlanta, Georgia, 1971.

City of Dallas, Texas. Industrial waste ordinance, 1969.

City of Fitchburg, Massachusetts. The discharge of waters and wastes into the public sewer system.

City of Houston, Texas. Disposal of industrial waste through city sewer system, 1974.

City of Muncie, Indiana. Muncie code of ordinances; Laws pertaining to this division, 1954–1967.

City of New York, New York. Rules and regulations relating to the use of the public sewer system for the discharge of sewage, industrial waste and other wastes, including surcharges and penalties. Amendment to the administrative code, section 687-1.0 industrial waste; sewer surcharges.

City of Olean, New York. Sewer use ordinance, September 1968.

City of Topeka, Kansas. Ordinance no. 13664, 1975

City of Wichita, Kansas. Title 16: Sewers, sewage disposal and drains, 1964. An ordinance amending sections of the code.

City of Wilmington, Delaware. Exclusion of materials detrimental to the sewerage system.

Commission of Jefferson County, Jefferson County, Alabama. Rules and regulations for discharge of waste into sanitary sewage system, April 1970. Resolution for industrial waste surcharge, September 1972.

County of Onondaga, Syracuse, New York. Rules and regulations relating to the use of the public sewer system, 1972.

East Bay Municipal Utility District, Oakland, California. Ordinance no. 27: Waste water control ordinance, 1972. Wastewater discharge permit parts A–G.

Elkin, H. F., et al. Effluent guidelines—Industry's point of view. *Pollution Engineering,* vol. 20, no. 6, p. 18, November–December 1974.

Environmental improvement agency of New Mexico, Santa Fe, New Mexico. Industrial waste ordinance (a model ordinance).

Escher, D. E. and A. J. Kicinski. Pretreatment requirements for industrial waste discharged to municipal treatment systems. Paper presented at the ASCE-EED Specialty Conference on Environmental Engineering Research, Development and Design, Pennsylvania State University, July 1974.

Finch, J. Consents and agreements. *Discharge of Industrial Effluents to Municipal Sewerage Systems,* p. 23. Proceedings of a Symposium of The Institute of Water Pollution Control, London, November 29–30, 1971.

Fisher, N. S. Present industrial effluent legislation and its shortcomings. *Discharge of Industrial Effluents to Municipal Sewerage Systems,* p. 14, Proceedings of a Symposium of the Institute of Water Pollution Control, London, November 29–30, 1971.

Gutierrez, A. F. Regulations and service charges for the treatment of industrial wastewater in federally assisted public facilities. Paper presented to the Southeast Section Convention of the American Waterworks Association, San Antonio, Texas, October 11, 1971.

Hamlin, W. G. Factors in the development of an industrial waste ordinance. *Proceedings of the 9th Industrial Waste Conference,* Purdue University, p. 14, 1954.

Lue-Hing, C. and E. W. Knight. Chicago vs. industry polluters. *Water and Wastes Engineering,* p. 71, September 1973.

Lavin, A. Energetic enforcement of industrial waste ordinances. *Proceedings of the 23rd Industrial Waste Conference,* Purdue University, p. 550, 1968.

The Metropolitan St. Louis Sewer District, St. Louis, Missouri. Ordinance no. 2289, May 1972. Ordinance no. 2412, March 1973. Ordinance no. 2444, June 1973.

The Metropolitan Sanitary District of Greater Chicago, Chicago, Illinois. Sewage and waste control ordinance as amended, 1972. Sewer permit ordinance, 1969, amended, 1972. Industrial waste division procedural manual.

Metropolitan Sewer Board, St. Paul, Minnesota. Sewage and waste control rules and regulations for the metropolitan disposal system, December 1, 1971.

Metropolitan Sewer District of Greater Cincinnati, Cincinnati, Ohio. Rules and regulations, December 4, 1968.

Metropolitan Sewerage System, Seattle, Washington. Resolution No. 2158: Regarding the control and disposal of industrial waste into the metropolitan sewerage system, July 1974.

Reich, J. S. Some experiences in the pretreatment of industrial waste going to the municipal sewer system of Philadelphia. *Proceedings of the 10th Industrial Waste* Conference, Purdue University, p. 244, 1955.

The Sanitary District of Rockford, Illinois. Ordinance no. 309: Pollutant discharge control ordinance of the sanitary district of Rockford, 1974.

Sanitation Districts of Los Angeles County, Los Angeles, California. An ordinance regulating sewer construction, sewer use and industrial wastewater discharges, April 1972. Instructions for obtaining a permit for industrial wastewater discharge. Instructions for filing an industrial wastewater treatment surcharge statement. Industrial wastewater charge rates, 1971. Technical report—Waste discharge to the ocean.

Sewer Utility of the city of Boulder, Boulder, Colorado. Ordinance no. 3836.

Shaw, R. E. Experience with waste ordinance and surcharges at Greensboro, North Carolina. *Journal of the Water Pollution Control Federation,* vol. 42, no. 1, p. 44, January 1970.

Simpson, J. R. Technical bases for assessing the strength, charges for treatment and treatability of trade wastes. *Water Pollution Control,* vol. 66, no. 2, p. 165, 1967.

State of Massachusetts. Suggested rules and regulations regarding the use of common sewers, 1974.

State of Vermont. Suggested model sewer use ordinance, January 1975.

Taylor, D. M. Establishing industrial waste ordinances. *Proceedings of the 10th Industrial Waste Conference,* Purdue University, p. 255, 1955.

Texas Water Quality Board, Austin, Texas. A suggested industrial waste ordinance, 1974.

Township of Towamencin, Pennsylvania. Rates, rules and regulations, April 1967.

Westchester County Environmental Facilities, Westchester County, New York. Sewer ordinance no. 1: Rules, regulations and ordinances governing the discharge of sewage, industrial wastes or other wastes.

MONITORING AND REPORTING ELEMENTS
OF A MONITORING PROGRAM

Albany, N. Y. *Industrial Wastewater Discharges.* Bureau of Water and Wastewater Utilities Management, Division of Pure Waters, June, 1969.

Available from the Health Education Service, P. O. Box 7283, Albany, N. Y. 12224.

Churchill, R. J. and T. A. Helbig. Monitoring wastewater? Try these methods. *Industrial Wastes,* p. 26, September/October 1974.

Permit Program Guidance for Self-Monitoring and Reporting Requirements. USEPA, Office of Water Enforcement, October 1, 1973.

Pojasek, R. B. NPDES permits and water analysis. *Environmental Science and Technology,* vol. 9, no. 4, p. 320, April 1975.

Quantitative Methods for Preliminary Design of Water Quality Surveillance Systems. USEPA, NTIS no. PB 219/010, November 1972.

Schafer, C. J. and N. Lailas. Complying with discharge regulations. *Environmental Science and Technology,* vol. 8, no. 10, p. 903, October 1974.

Stack, Vernon T., Jr. Water quality monitoring must be action-oriented. *Water and Waste Engineering,* vol. 8, no. 3, p. 310, March 1971.

Wrigley, K. J. and F. Ashworth. The need for, and methods of, monitoring and control of industrial discharges to sewers. *Discharge of Industrial Effluents to Municipal Sewerage Systems,* p. 91. Proceedings of the Symposium of the Institute of Water Pollution Control, London, November 29–30, 1971.

Zabban, W. In-process monitoring. Paper presented at the EPA Technology Transfer Seminar on Monitoring Industrial Wastewater, Arlington, Virginia, January 9, 1975.

FIELD CONSIDERATIONS

Babcock, R. H. Industrial waste treatment plant instrumentation. *Water and Waste Engineering,* vol. 5, no. 9, p. 3, September 1968.

Babcock, R. H. Evaluation of instrumentation and control. *Journal of the Water Pollution Control Federation,* vol. 44, no. 7, p. 1416, July 1972.

Beach, M. I. and J. S. Beach, Jr. A new technique for industrial waste sampling. *Industrial Wastes,* p. 28, January/February 1973.

Beach, M. I., and C. F. Gurnham. The role of automatic sampling in industrial waste control. *Mid Atlantic Industrial Waste Conference,* no. 5, p. 225, 1971.

Brailsford, H. D. A new automatic sampler for industrial outfall, streams and sewers. *Water and Sewage Works,* September 1968.

Craddock, J. M. Make water pollution control a meaningful local responsibility. *The American City,* p. 63, May 1974.

Evans, M. R. and R. Edgar. Automatic sampling and measurement of small liquid flows. *Water Pollution Control,* vol. 70, 1971.

Forester, R. and D. Overland. Portable device to measure industrial wastewater flow. *Journal of the Water Pollution Control Federation,* vol. 46, no. 4, p. 777, April 1974.

Hom, L. W. Remote sensing of water pollution. *Journal of the Water Pollution Control Federation,* vol. 40, no. 10, p. 1728, October 1968.

Jones, R. H. Instrumentation for water pollution control. *Pollution Engineering,* vol. 3, no. 6, p. 22, November/December 1971.

Klein, L. A. and A. Montague. Gauging and sampling industrial wastewater (open channel). *Journal of the Water Pollution Control Federation,* vol. 42, no. 8, p. 1468, August 1970.

Levin, V. H., and A. Latten. Automatic samplers for sewage and effluents. *Process Biochemistry,* p. 15, June 1973.

Lysyj, I., et al. Monitoring industrial pollutants by pyrolysis—Methane detection method. *Journal of the Water Pollution Control Federation,* vol. 40, no. 5, part 2, p. R181, May 1968.

Maylath, R. E. Monitoring New York's water automatically. *Journal of the American Water Works Association,* vol. 63, p. 517, August 1971.

McRae, A. D., et al. Detection and monitoring of phenolic wastewater. *Proceedings of the 14th Industrial Waste Conference,* Purdue University, 1959.

Medved, T. K., et al. Instrumentation for measurement of wastewater flow. *Journal of the Water Pollution Control Federation,* vol. 44, no. 5, p. 820, May 1972.

Nichols, P. R. When you go into a manhole or a sewer, you should understand sewer gases. *Deeds & Data,* p. 2, January 1975.

O'Brien, J. E. and R. A. Olsen. Evaluation of an automatic chemical analysis monitor for water quality parameters. *Journal of the Water Pollution Control Federation,* vol. 42, no. 3, p. 380, March 1970.

Ostendorf, R. G. and J. F. Byrd. Modern monitoring of a treated industrial effluent. *Journal of the Water Pollution Control Federation,* vol. 41, no. 1, p. 89, January 1969.

Rittmiller, L. A., et al. Comparison of air and water pollution instrumentation. *Pollution Engineering,* vol. 3, no. 6, p. 26, November–December 1971.

Salvatorelli, J. Value of instrumentation in wastewater treatment. *Journal of the Water Pollution Control Federation,* vol. 40, no. 1, p. 101, January 1968.

Sampling and monitoring feature. *Water and Waste Treatment,* vol. 16, no. 10, p. 11, October 1973.

Snowden, F. C. Instrumentation in pollution control. *Industrial Water Engineering,* vol. 7, no. 6, p. 22, June 1970.

Tarazi, D. S., et al. Comparison of wastewater sampling techniques. *Journal of the Water Pollution Control Federation,* vol. 42, no. 5, p. 708, May 1970.

Thorsen, T. and R. Oen. How to measure industrial wastewater flow. *Chemical Engineering,* vol. 82, no. 4, p. 95, February 17, 1975.

Vought, J. H. Monitoring and treatment of cyanide-bearing plating wastes. *Journal of the Water Pollution Control Federation,* vol. 39, no. 12, p. 1971, December 1967.

Ward, R. C. Routine surveillance alternatives for water quality management. *Journal of the Water Pollution Control Federation,* vol. 46, no. 12, p. 2645, December 1974.

Wood, L. B., and H. H. Stanbridge. Automatic samplers. *Water Pollution Control,* vol. 67, p. 495, 1968.

LABORATORY CONSIDERATIONS

Andelman, J. B. et al. Determination of organics in water. *Proceedings of the 20th Industrial Waste Conference,* Purdue University, p. 220, 1965.

Arin, M. L. Monitoring with carbon analyzers. *Environmental Science and Technology,* vol. 8, no. 10, p. 898, October 1974.

Arthur, R. M. An automated BOD respirometer. *Proceedings of the 19th Industrial Waste Conference,* Purdue University, p. 628, 1964.

Analysis for Mercury in Water, A Preliminary Study of Methods. USEPA no. R4-72-003, September 1972.

Atanus, H. TLC finds hexane solubles. *Water and Wastes Engineering,* vol. 11, no. 10, p. 26, October 1974.

Atomic absorption spectrophotometer facilitates water analysis. *Water and Sewage Works,* vol. 121, no. 1, p. 27, January 1974.

Chapman, B., et al. Automated analysis: The determination of ammoniacal, nitrous and nitric nitrogen in river waters, sewage effluents and trade effluents. *Water Pollution Control,* vol. 66, no. 2, p. 185, 1967.

Babcock, R. H. Ion-selective electrodes for quality measurement and control. *Journal of the American Water Works Association,* p. 26, January 1975.

Bailey, A. R. Thin layer chromatography as a sorting test for metals in trade effluent. *Water Pollution Control,* vol. 68, no. 4, p. 449, 1969.

Baker, R. A. and M.-D. Luh. Mercury analysis and toxicity: A review. *Water and Sewage Works,* vol. 118, no. 5, p. IW-21, May 1971. (Also included in *Industrial Wastes,* May/June 1971.)

Banerji, S. K. Laboratory tests for plant operation control and stream quality measurement. *Journal of the Water Pollution Control Federation,* vol. 43, no. 3, p. 399, March 1971.

Blackmore, R. H. and D. Voshel. Rapid determination of total organic carbon (TOC) in sewage. *Water and Sewage Works,* vol. 114, no. 10, p. 398, October 1967.

Bramer, H. C., et al. Instrument for monitoring trace organic compounds in water. *Water and Sewage Works,* vol. 113, no. 8, p. 275, August 1966.

Brown, J. A., Jr. A rapid wastewater sensitivity test. *Industrial Waste,* p. 28, May/June 1972.

Chanin, G., et al. A safe solvent for oil and grease analyses. *Journal of the Water Pollution Control Federation,* vol. 39, no. 11, p. 1892, November 1967.

Cheremisinoff, P. N., and Y. H. Habib. Cadmium, chromium, lead, mercury: A plenary account for water pollution, part I—Occurrence, toxicity and detection. *Sewage Works,* vol. 119, no. 7, p. 73, July 1972.

Cheremisinoff, P. N., and R. Young. 1975 annual review of the new developments in water quality instrumentation. *Pollution Engineering,* p. 28, March 1975.

Clark, H. A. Characterization of industrial wastes by instrumental analysis. *Proceedings of the 23rd Industrial Waste Conference,* Purdue University, p. 26, 1967.

Cochran, L. G. and F. D. Bess. Waste monitoring by gas chromatography. *Journal of the Water Pollution Control Federation,* vol. 38, no. 12, p. 2002, December 1966.

Cramer, C. D. *Detection and Characterization of Animal/Vegetable and Petroleum Oil in Municipal Wastewater by Thin Layer Chromatography.* Oak Brook, Illinois: Nader Chemical Co.

Cunningham, M. F., et al. The work of the Dalmarnock Laboratory, Glasgow. *Water Pollution Control,* vol. 72, no. 4, p. 392, 1973.

Current Practice in GC–MS Analysis of Organics in Water. USEPA, NTIS no. PB 224 947, August 1973.

Determination of heavy metals in municipal sewage plant sludges by neutron activation analysis. *Water, Air and Soil Pollution,* vol. 3, no. 3, p. 327, September 1974.

Ellison, W. K., and T. E. Wallbank. Solvents in sewage and industrial waste waters: Identification and determination. *Water Pollution Control,* vol. 73, no. 6, p. 656, 1974.

Environmental Applications of Advanced Instrumental Analyses: Assistance Projects, FY 69–71. USEPA, May 1973.

Estimation of Polychlorinated Biphenyls in the Presence of DDT-Type Compounds. USEPA, NTIS no. PB 233 599, June 1974.

Ettinger, M. B., and R. C. Kroner. The determination of phenolic materials in industrial wastes. *Proceedings of the 5th Industiral Waste Conference,* Purdue University, p. 345, 1949.

Evans, R. L., et al. Mercury in public sewer systems. *Water and Sewage Works,* p. 74, February 1973.

Field Tests of LAS Biodegradability. The Soap and Detergent Association, Scientific and Technical Report no. 2, September 1965.

Fluorescent Probes in the Detection of Insecticides in Water. USEPA, NTIS no. PB 221 336, April 1973.

Ford, D. L. Application of the total carbon analyzer for industrial wastewater evaluation. *Proceedings of the 23rd Industrial Waste Conference,* Purdue University, p. 989, 1968.

Frant, M. S. Detecting pollutants with chemical-sensing electrodes. *Environmental Science and Technology,* vol. 8, no. 3, p. 224, March 1974.

Gales, M. E., Jr., and R. Booth. *Simultaneous and Automated Determination of Total Phosporus and Total Kjeldahl Nitrogen.* USEPA, NTIS no. PB 232 710, p. 19, May 1974.

Gaudy, A. F. and M. Ramanathan. A colorimetric method for determining chemical oxygen demand. *Proceedings of the 19th Industrial Waste Conference,* Purdue University, p. 915, 1964.

Gupta, K. B. and A. E. Zanoni. Total phosphorus analysis: Persulfate on ashing? *Water and Sewage Works,* vol. 121, no. 7, p. 74, July 1974.

Guyon, J. C. and W. D. Shults. Rapid phosphate determination by fluorimetry. *Journal of the American Water Works Association,* vol. 63, p. 403, August 1969.

Henry, L. *Thin Layer Chromatographic Method for the Determination of*

Petroleum on Mineral Hydrocarbons and Other Natural or Synthetic Oils and Greases. Chagrin Falls, Ohio: Mogul Corporation.

Hewitt, P. J., and H. B. Austin. Determination of cyanide in industrial effluents. *Water Pollution Control,* vol. 71, no. 4, p. 381, 1972.

Hwang, C. P. and C. R. Forsberg. Polarographic method for nitrate and dissolved oxygen analysis. *Water and Sewage Works,* vol. 120, no. 4, p. 71, April 1973.

Kalinske, A. A., et al. Cobalt interference in the non–steady state clean water test. *Water and Sewage Works,* vol. 120, no. 7, p. 54, July 1973.

Katz, S., et al. The determination of stable organic compounds in waste effluents at microgram per liter levels by automatic high-resolution ion exchange chromatography. *Water Research,* vol. 6, no. 9, p. 1029, September 1972.

Kawahara, F. K., et al. Thin layer and gas chromatographic analysis of parathion and methyl parathion in the presence of chlorinated hydro-carbons. *Journal of the Water Pollution Control Federation,* vol. 39, no. 3, p. 446, March 1967.

Kerber, J. D. Detection of trace metals in water. *Industrial Water Engineering,* vol. 10, no. 5, September/October 1973.

Kopp, J. F., et al. "Cold vapor" method for determining mercury. *Journal of the American Water Works Association,* vol. 64, p. 20, January 1972.

Larson, T. E., et al. The determination of total organic carbon in water. *Proceedings of the 19th Industrial Waste Conference,* Purdue University, p. 762, 1964.

Lee, E. G. H., and C. C. Walden. A rapid method for the estimation of trace amounts of kerosene in effluents. *Water Research,* vol. 4, no. 9, p. 641, 1970.

Literature Survey of Instrumental Measurements of Biochemical Oxygen Demand for Control Application, 1960–1973. Cincinnati, Ohio: National Environmental Research Center, Office of Research Development, USEPA, Environmental Monitoring Series, EPA-670/4-74-001, February 1974.

Lively, L., et al. Identification of petroleum products in water. *Proceedings of the 20th Industrial Waste Conference,* Purdue University, p. 657, 1965.

Lysyj, I., et al. Rapid instrumental measurement of the organic load in wastewaters. *Journal of the Water Pollution Control Federation,* vol. 41, no. 5, p. 831, May 1969.

Maeller, C. Z., et al. Differentiation of LAS and ABS in water. *Journal of the Water Pollution Control Federation,* vol. 39, no. 10, part 2, p. R92, October 1967.

Maier, W. J. and H. L. McConnell. Carbon measurements in water quality monitoring. *Journal of the Water Pollution Control Federation,* vol. 46, no. 4, p. 623, April 1974.

Malhotra, S. K. and A. E. Zanoni. Chloride interference in nitrate nitrogen determination. *Journal of the American Water Works Association,* vol. 62, no. 9, p. 568, September 1970.

McFarren, E. F. New, simplified methods for metal analysis. *Journal of the American Water Works Association,* vol. 64, p. 28, January 1972.

McFarren, E. F. and R. J. Lishka. The use of collaborative studies to evaluate water analysis instruments. *Journal of the Water Pollution Federation,* vol. 43, p. 67, January 1971.

McKeown, J. J., et al. Comparative studies of dissolved oxygen analysis methods. *Journal of the Water Pollution Control Federation,* vol. 39, no. 8, p. 1323, August 1967.

Mentink, A. *Instrumentation for Water Quality Determination.* ASCE, Water Resources Engineering Conference, March 8–12, 1965.

Metals in sewage measured simply but accurately. *The American City,* p. 40, August, 1972.

Molof, A. H. and N. S. Zaleiko. The detection of organic pollution by automated COD. *Proceedings of the 19th Industrial Waste Conference,* Purdue University, p. 540, 1964.

Mullis, M. K. and E. D. Schroeder. A rapid biochemical oxygen demand test suitable for operational control. *Journal of the Water Pollution Control Federation,* vol. 43, no. 2, p. 209, February 1971.

Musselwhite, C. C. and K. W. Petts. An automated method for the determination of formaldehyde in sewage and sewage effluents. *Water Pollution Control,* vol. 73, no. 4, p. 443, 1974.

Nelson, K. H., and I. Lysyj. Analysis of water for molecular hydrogen cyanide. *Journal of the Water Pollution Control Federation,* vol. 43, no. 5, p. 799, May 1971.

Offner, H. G. and E. F. Witucki. Toxic inorganic materials and their emergency detection by the polarographic method. *Journal of the American Water Works Association,* vol. 60, no. 8, p. 947, August 1968.

Organic Pollutant Identification Utilizing Mass Spectrometry. USEPA, NTIS no. PB 224 544, July 1973.

Performance of the Union Carbide Dissolved Oxygen Analyzer. Cincinnati, Ohio: Office of Research and Development, USEPA, Environmental Monitoring Series, EPA 670/4-73-018, July 1973.

Porter, J. D. and W. W. Sanderson. Polarographic scanning of industrial waste samples. *Proceedings of the 9th Industrial Waste Conference,* Purdue University, 1954.

Prager, M. *Automated water monitoring instrument for phosphorus contents.* USEPA, NTIS no. PB 222 772, June 1973.

Pyrographic Gross Characterization of Water Contaminants. USEPA no. EPA R2-73-227, May 1973.

A Quick Biochemical Oxygen Demand Test. USEPA, Water Pollution Control Research Series, EPA no. 16050 EMF (06171).

Reynolds, J. F. Comparison studies of Winkler vs. oxygen sensor. *Journal of the Water Pollution Control Federation,* vol. 41, no. 12, p. 2002, December 1969.

Reynolds, J. F. and K. A. Goellner. Statistical evaluation of BOD versus ODI. *Water and Sewage Works,* vol. 121, no. 1, p. 31, January 1974.

Riehl, M. L. and E. G. Will. Analytical determination of metals affecting sewage treatment. *Proceedings of the 4th Industrial Waste Conference, Purdue University,* 1948.

Roesler, J. F. and R. H. Wise. Variables to be measured in wastewater treatment plant monitoring and control. *Journal of the Water Pollution Control Federation,* vol. 46, no. 7, p. 1769, July 1974.

Self-contained sampling and measurement system features respirometer. *Water and Sewage Works,* vol. 121, no. 2, p. 53, February 1974.

Stenger, V. A., and C. E. Van Hall. Analysis of municipal and chemical wastewaters by an instrumental method for COD determination. *Journal of the Water Pollution Control Federation,* vol. 40, no. 10, p. 1755, October 1968.

Sugar, W. J. and R. A. Conway. Gas–liquid chromatographic techniques for petrochemical wastewater analysis. *Journal of the Water Pollution Control Federation,* vol. 40, no. 9, p. 1622, September 1968.

Test Procedure and Standards—ABS and LAS Biodegradability. The Soap and Detergent Association, Scientific and Technical Report no. 3, January 1966.

Thruston, A. D. Jr. A fluorometric method for the determination of lignin sulfonates in natural waters. *Journal of the Water Pollution Control Federation,* vol. 42, no. 8, p. 1551, August 1970.

Van Loon, J. C., et al. The determination of heavy metals in domestic sewage treatment plant wastes. *Water, Air and Soil Pollution,* vol. 2, no. 4, p. 473, December 1973.

Vought, J. H. Monitoring and treatment of cyanide-bearing plating wastes. *Journal of the Water Pollution Control Federation,* vol. 39, no. 12, p. 1971, December 1967.

Willey, B. F., et al. Atomic absorption spectrophotometry simplifies heavy-metals analysis. *Journal of the American Water Works Association,* vol. 64, p. 303, May 1972.

Williamson, T., and A. S. Millar. Instrumentation in water pollution control analysis. *Water Pollution Control,* vol. 70, 1971.

Woods, C. Determination of proteins in waste water. *Process Industrial Waste Control,* vol. 49, no. 4, p. 501, July 1965.

USER CHARGE SYSTEMS

Development of a State Effluent Charge system. USEPA, NTIS no. PB 210 711, February 1972.

Harkness, N. Methods of charging for the reception, treatment and disposal of toxic wastes. *Water Pollution Control,* vol. 69, 1970.

Lewin, V. H. Industrial effluents: Problems of recovering costs. *Discharge of Industrial Effluents to Municipal Sewerage Systems,* p. 77. Proceedings of a Symposium of the Institute of Water Pollution Control, London, November 29–30, 1971.

Seagraves, J. A. Industrial waste charges. *Journal of the Environmental Engineering Division, ASCE,* vol. 99, no. EE 6, p. 873, December 1973.

Simpson, J. R., and G. A. Truesdale. Methods of charging for the treatment and disposal of industrial effluents in municipal sewerage systems. *Discharge of Industrial Effluents to Municipal Sewerage Systems,* p. 65. Proceedings of a Symposium of the Institute of Water Pollution Control, London, November 29–30, 1971.

COMPLIANCE PROCEDURES

Local Pretreatment Program Requirements and Guidance. Springfield, Virginia: Environmental Technology Consultants, Inc., June 1979.

APPENDIX
Model Ordinances

WATER POLLUTION CONTROL
FEDERATION MODEL ORDINANCE (1975)

The . . . suggested ordinance . . . is not intended to be adapted to any specific use merely by filling in the inserts. Some sections may be applicable directly as written. Others must be modified to suit conditions that exist and can reasonably be anticipated.

All items obviously requiring local adaptation such as names, type of municipal corporation and/or regulatory agency, and titles of officials and those that should be considered with particular care before acceptance such as limiting values and figures are marked in the suggested ordinance by enclosure in brackets.

In general, the suggested ordinance should not be taken verbatim . . . with only the brackets filled in as necessary. Rather, each article and each section should be critically reviewed in the light of peculiarities of each local situation and revised as necessary to meet the varying needs of each locality. . . .

The legal, technical, and practical aspects of each section should . . . be passed on by the variously qualified members of the ordinance committee, with due consideration to good general practice as well as to local conditions and requirements. No specific limits of biochemical oxygen demand (BOD) and solids concentration are included in the suggested ordinance.

When drafted to the satisfaction of the ordinance committee, it is suggested that a review copy be submitted to the state and federal water pollution control authority. Such procedure will reveal possible conflicts with state and federal requirements and will yield the benefits of independent, competent, and impartial criticism. The preliminary draft should also be submitted for review to the consulting sanitary engineer of the municipality and/or sewer authority if he is not a member of the drafting committee. Copies of the ordinance as finally adopted should always be filed with the state water pollution control agency.

Suggested application forms for the permits referred to in the suggested ordinance are presented [at the end].

The Suggested Ordinance

An ordinance regulating the use of public and private sewers and drains, private wastewater disposal, the installation and connection of building sewers, and the discharge of waters and wastes into the public sewer system(s): and providing penalties for violations thereof: in the city of ____ _______________________________, county of _______________________________, state of ____.

Be it ordained and enacted by the council of the city of _______________________, state of _______________________________________ as follows:

ARTICLE I

*Definitions**

Unless the context specifically indicates otherwise, the meaning of terms used in this ordinance shall be as follows:

SEC. 1. Biochemical oxygen demand (BOD) shall mean the quantity of oxygen utilized in the biochemical oxidation of organic matter under standard laboratory procedure in five (5) days at 20°C, expressed in milligrams per liter.

SEC. 2. "Building drain" shall mean that part of the lowest horizontal piping of a drainage system which receives the discharge from soil, waste, and other drainage pipes inside the walls of the building and conveys it to the building sewer, beginning five (5) feet (1.5 meters) outside the inner face of the building wall.

SEC. 3. "Building sewer" shall mean the extension from the building drain to the public sewer or other place of disposal, also called house connection.

SEC. 4. "Combined sewer" shall mean a sewer intended to receive both wastewater and storm or surface water.

SEC. 5. "Easement" shall mean an acquired legal right for the specific use of land owned by others.

SEC. 6. "Floatable oil" is oil, fat, or grease in a physical state such that it will separate by gravity from wastewater by treatment in an approved pretreatment facility. A wastewater shall be considered free of floatable fat if it is properly pretreated and the wastewater does not interfere with the collection system.

SEC. 7. "Garbage" shall mean the animal and vegetable waste resulting from the handling, preparation, cooking, and serving of foods.

SEC. 8. "Industrial wastes" shall mean the wastewater from industrial processes, trade, or business as distinct from domestic or sanitary wastes.

SEC. 9. "Natural outlet" shall mean any outlet, including storm sewers and combined sewer overflows, into a watercourse, pond, ditch, lake, or other body of surface or groundwater.

SEC. 10. "May" is permissive (see "shall," Sec. 18).

SEC. 11. "Person" shall mean any individual, firm, company, association, society, corporation, or group.

SEC. 12. "pH" shall mean the logarithm of the reciprocal of the hydrogen-ion concentration. The concentration is the weight of hydrogen ions, in grams, per liter of solution. Neutral water, for example, has a pH value of 7 and a hydrogen-ion concentration of 10^{-7}.

SEC. 13. "Properly shredded garbage" shall mean the wastes from the preparation, cooking, and dispensing of food that have been shredded to such a degree that all particles will be carried freely under the flow conditions normally prevailing in public sewers, with no particle greater than ½ inch (1.27 centimeters) in any dimension.

---

*Included among the definitions of a particular ordinance should be those for such terms as "city," "village," and various officials as are locally applicable.

SEC. 14. "Public sewer" shall mean a common sewer controlled by a governmental agency or public utility.

SEC. 15. "Sanitary sewer" shall mean a sewer that carries liquid and water-carried wastes from residences, commercial buildings, industrial plants, and institutions together with minor quantities of ground, storm, and surface waters that are not admitted intentionally.

SEC. 16. "Sewage" is the spent water of a community. The preferred term is "wastewater," Sec. 24.

SEC. 17. "Sewer" shall mean a pipe or conduit that carries wastewater or drainage water.

SEC. 18. "Shall" is mandatory (see "may," Sec. 10).

SEC. 19. "Slug" shall mean any discharge of water or wastewater which in concentration of any given constituent or in quantity of flow exceeds for any period of duration longer than fifteen (15) minutes more than five (5) times the average twenty-four (24) hour concentration or flows during normal operation and shall adversely affect the collection system and/or performance of the wastewater treatment works.

SEC. 20. "Storm drain" (sometimes termed "storm sewer") shall mean a drain or sewer for conveying water, groundwater, subsurface water, or unpolluted water from any source.

SEC. 21. "Superintendent" shall mean the [superintendent of wastewater facilities, and/or of wastewater treatment works, and/or of water pollution control] of the [city] of [   ], or his authorized deputy, agent, or representative.

SEC. 22. "Suspended solids" shall mean total suspended matter that either floats on the surface of, or is in suspension in, water, wastewater, or other liquids, and that is removable by laboratory filtering as prescribed in "Standard Methods for the Examination of Water and Wastewater" and referred to as nonfilterable residue.

SEC. 23. "Unpolluted water" is water of quality equal to or better than the effluent criteria in effect or water that would not cause violation of receiving water quality standards and would not be benefited by discharge to the sanitary sewers and waste-water treatment facilities provided.

SEC. 24. "Wastewater" shall mean the spent water of a community. From the standpoint of source, it may be a combination of the liquid and water-carried wastes from residences, commercial buildings, industrial plants, and institutions, together with any groundwater, surface water, and stormwater that may be present.

SEC. 25. "Wastewater facilities" shall mean the structures, equipment, and processes required to collect, carry away, and treat domestic and industrial wastes and dispose of the effluent.

SEC. 26. "Wastewater treatment works" shall mean an arrangement of devices and structures for treating wastewater, industrial wastes, and sludge. Sometimes used as synonymous with "waste treatment plant" or "wastewater treatment plant" or "water pollution control plant."

SEC. 27. "Watercourse" shall mean a natural or artificial channel for the passage of water either continuously or intermittently.

SEC. 28. "Hearing board" shall mean that board appointed according to the provision of Article [   ]. (This section to be included only if optional article entitled "hearing boards" is made a part of the ordinance.)

ARTICLE II

Use of Public Sewers Required

SEC. 1. It shall be unlawful for any person to place, deposit, or permit to be deposited in any unsanitary manner on public or private property within the [city] of [   ], or in any area under the jurisdiction of said [city], any human or animal excrement, garbage, or objectionable waste.

SEC. 2. It shall be unlawful to discharge to any natural outlet within the [city] of [   ], or in any area under the jurisdiction of said [city], any wastewater or other polluted waters, except where suitable treatment has been provided in accordance with subsequent provisions of this ordinance.

SEC. 3. Except as hereinafter provided, it shall be unlawful to construct or maintain any privy, privy vault, septic tank, cesspool, or other facility intended or used for the disposal of wastewater.

SEC. 4. The owner(s) of all houses, buildings, or properties used for human occupancy, employment, recreation, or other purposes, situated within the [city] and abutting on any street, alley, or right-of-way in which there is now located or may in the future be located a public sanitary or combined sewer of the [city], is hereby required at the owner(s)' expense to install suitable toilet facilities therein, and to connect such facilities directly with the proper public sewer in accordance with the provisions of this ordinance, within [ninety (90) days] after date of official notice to do so, provided that said public sewer is within [one hundred (100) feet (30.5 meters)] of the property line.

ARTICLE III

Private Wastewater Disposal

SEC. 1. Where a public sanitary or combined sewer is not available under the provisions of Article II, Section 4, the building sewer shall be connected to a private wastewater disposal system complying with the provisions of this article.

SEC. 2. Before commencement of construction of a private wastewater disposal system the owner(s) shall first obtain a written permit signed by the [superintendent]. The application for such permit shall be made on a form furnished by the [city], which the applicant shall supplement by any plans, specifications, and other information as are deemed necessary by the [superintendent]. A permit and inspection fee of [   ] dollars shall be paid to the [city] at the time the application is filed. . . .

SEC. 3. A permit for a private wastewater disposal system shall not become effective until the installation is completed to the satisfaction of the [superintendent]. The [superintendent] shall be allowed to inspect the work at any stage of construction, and, in any event, the applicant for the permit shall notify the [superintendent] when the work is ready for final inspection, and before any underground portions are covered. The inspection shall be made within [   ] hours of the receipt of notice by the [superintendent].

SEC. 4. The type, capacities, location, and layout of a private wastewater disposal system shall comply with all recommendations of the department of public health of the state of [   ]. No permit shall be issued for any private wastewater disposal system employing subsurface soil absorption facilities where the area of the lot is less than [   ] square feet [square meters]. No septic tank or cesspool shall be permitted to discharge to any natural outlet.

SEC. 5. At such time as a public sewer becomes available to a property served by a private wastewater disposal system, as provided in Article III, Section 4, a direct connection shall be made to the public sewer within sixty (60) days in compliance with this ordinance, and any septic tanks, cesspools, and similar private wastewater disposal facilities shall be cleaned of sludge and filled with suitable material.

SEC. 6. The owner(s) shall operate and maintain the private wastewater disposal facilities in a sanitary manner at all times, at no expense to the [city]. (Some communities may desire to provide additional requirements that sludge removal from private disposal systems be performed by licensed operators and disposed of in some particular fashion.)

SEC. 7. No statement contained in this article shall be construed to interfere with any additional requirements that may be imposed by the health officer.

ARTICLE IV

Building Sewers and Connections

SEC. 1. No unauthorized person(s) shall uncover, make any connections with or opening into, use, alter, or disturb any public sewer or appurtenance thereof without first obtaining a written permit from the [superintendent].

SEC. 2. There shall be two (2) classes of building sewer permits: (*a*) for residential and commercial service, and (*b*) for service to establishments producing industrial wastes. In either case, the owner(s) or his agent shall make application on a special form furnished by the [city]. The permit application shall be supplemented by any plans, specifications, or other information considered pertinent in the judgment of the [superintendent]. A permit and inspection fee of [   ] dollars for a residential or commercial building sewer permit and [   ] dollars for an industrial building sewer permit shall be paid to the [city] at the time the application is filed.

SEC. 3. All costs and expenses incidental to the installation and connection of the building sewer shall be borne by the owner(s). The owner(s) shall indemnify the [city] from any loss or damage that may directly or indirectly be occasioned by the installation of the building sewer.

SEC. 4. A separate and independent building sewer shall be provided for every building; except where one building stands at the rear of another on an interior lot and no private sewer is available or can be constructed to the rear building through an adjoining alley, court, yard, or driveway, the front building may be extended to the rear building and the whole considered as one building sewer, but the [city] does not and will not assume any obligation or responsibility for damage caused by or resulting from any such single connection aforementioned.

SEC. 5. Old building sewers may be used in connection with new buildings only when they are found, on examination and test by the [superintendent], to meet all requirements of this ordinance.

Sec. 6. The size, slope, alignment, materials of construction of a building sewer, and the methods to be used in excavating, placing of the pipe, jointing, testing, and backfilling the trench, shall all conform to the requirements of the building and plumbing code or other applicable rules and regulations of the [city]. In the absence of code provisions or in amplification thereof, the materials and procedures set forth in appropriate specifications of the ASTM and WPCF Manual of Practice No. 9 shall apply.

Sec. 7. Whenever possible, the building sewer shall be brought to the building at an elevation below the basement floor. In all buildings in which any building drain is too low to permit gravity flow to the public sewer, sanitary sewage carried by such building drain shall be lifted by an approved means and discharged to the building sewer.

Sec. 8. No person(s) shall make connection of roof downspouts, foundation drains, areaway drains, or other sources of surface runoff or groundwater to a building sewer or building drain which in turn is connected directly or indirectly to a public sanitary sewer unless such connection is approved by the superintendent for purposes of disposal of polluted surface drainage.

Sec. 9. The connection of the building sewer into the public sewer shall conform to the requirements of the building and plumbing code or other applicable rules and regulations of the [city], or the procedures set forth in appropriate specifications of the ASTM and the WPCF Manual of Practice No. 9. All such connections shall be made gastight and watertight and verified by proper testing. Any deviation from the prescribed procedures and materials must be approved by the [superintendent] before installation.

Sec. 10. The applicant for the building sewer permit shall notify the [superintendent] when the building sewer is ready for inspection and connection to the public sewer. The connection and testing shall be made under the supervision of the [superintendent] or his representative.

Sec. 11. All excavations for building sewer installation shall be adequately guarded with barricades and lights so as to protect the public from hazard. Streets, sidewalks, parkways, and other public property disturbed in the course of the work shall be restored in a manner satisfactory to the [city].

Article V

Use of the Public Sewers

Sec. 1. No person(s) shall discharge or cause to be discharged any unpolluted waters such as stormwater, groundwater, roof runoff, subsurface drainage, or cooling water to any sewer, except stormwater runoff from limited areas, which stormwater may be polluted at times, may be discharged to the sanitary sewer by permission of the [superintendent].

Sec. 2. Stormwater other than that exempted under Section 1, Article V, and all other unpolluted drainage shall be discharged to such sewers as are specifically designated as combined sewers or storm sewers or to a natural outlet approved by the [superintendent] and other regulatory agencies. Unpolluted industrial cooling water or process waters may be discharged, on approval of the [superintendent], to a storm sewer, combined sewer, or natural outlet.

SEC. 3. No person(s) shall discharge or cause to be discharged any of the following described waters or wastes to any public sewers:

(a) Any gasoline, benzene, naptha, fuel oil, or other flammable or explosive liquid, solid, or gas.

(b) Any waters containing toxic or poisonous solids, liquids, or gases in sufficient quantity, either singly or by interaction with other wastes, to injure or interfere with any waste treatment process, constitute a hazard to humans or animals, create a public nuisance, or create any hazard in the receiving waters of the wastewater treatment plant.

(c) Any waters or wastes having a pH lower than [5.5], or having any other corrosive property capable of causing damage or hazard to structures, equipment, and personnel of the wastewater works.

(d) Solid or viscous substances in quantities or of such size capable of causing obstruction to the flow in sewers, or other interference with the proper operation of the wastewater facilities such as, but not limited to, ashes, bones, cinders, sand, mud, straw, shavings, metal, glass, rags, feathers, tar, plastics, wood, unground garbage, whole blood, paunch manure, hair and fleshings, entrails, and paper dishes, cups, milk containers, etc., either whole or ground by garbage grinders.

SEC. 4. The following described substances, materials, waters, or waste shall be limited in discharges to municipal systems to concentrations or quantities which will not harm either the sewers, wastewater treatment process, or equipment, will not have an adverse effect on the receiving stream, or will not otherwise endanger lives, limb, public property, or constitute a nuisance. The [superintendent] may set limitations lower than the limitations established in the regulations below if in his opinion such more severe limitations are necessary to meet the above objectives. In forming his opinion as to the acceptability, the [superintendent] will give consideration to such factors as the quantity of subject waste in relation to flows and velocities in the sewers, materials of construction of the sewers, the wastewater treatment process employed, capacity of the wastewater treatment plant, degree of treatability of the waste in the wastewater treatment plant, and other pertinent factors. The limitations or restrictions on materials or characteristics of waste or wastewaters discharged to the sanitary sewer which shall not be violated without approval of the [superintendent] are as follows:

(a) Wastewater having a temperature higher than 150° Fahrenheit (65° Celsius).

(b) Wastewater containing more than 25 milligrams per liter of petroleum oil, nonbiodegradable cutting oils, or product of mineral oil origin.

(c) Wastewater from industrial plants containing floatable oils, fat, or grease.

(d) Any garbage that has not been properly shredded (see Article I, Section 13.) Garbage grinders may be connected to sanitary sewers from homes, hotels, institutions, restaurants, hospitals, catering establishments, or similar places where garbage originates from the preparation of food in kitchens for the purpose of consumption on the premises or when served by caterers.

(*e*) Any waters or wastes containing iron, chromium, copper, zinc, and similar objectionable or toxic substances to such degree that any such material received in the composite wastewater at the wastewater treatment works exceeds the limits established by the [superintendent] for such materials.

(*f*) Any waters or wastes containing odor-producing substances exceeding limits which may be established by the [superintendent].

(*g*) Any radioactive wastes or isotopes of such half-life or concentration as may exceed limits established by the [superintendent] in compliance with applicable state or federal regulations.

(*h*) Quantities of flow, concentrations, or both which constitute a "slug" as defined herein.

(*i*) Waters or wastes containing substances which are not amenable to treatment or reduction by the wastewater treatment processes employed, or are amenable to treatment only to such degree that the wastewater treatment plant effluent cannot meet the requirements of other agencies having jurisdiction over discharge to the receiving waters.

(*j*) Any water or wastes which, by interaction with other water or wastes in the public sewer system, release obnoxious gases, form suspended solids which interfere with the collection system, or create a condition deleterious to structures and treatment processes.

SEC. 5. If any waters or wastes are discharged or are proposed to be discharged to the public sewers, which waters contain the substances or possess the characteristics enumerated in Section 4 of this Article, and which in the judgment of the [superintendent], may have a deleterious effect upon the wastewater facilities, processes, equipment, or receiving waters, or which otherwise create a hazard to life or constitute a public nuisance, the [superintendent] may:

(*a*) Reject the wastes,

(*b*) Require pretreatment to an acceptable condition for discharge to the public sewers,

(*c*) Require control over the quantities and rates of discharge, and/or

(*d*) Require payment to cover added cost of handling and treating the wastes not covered by existing taxes or sewer charges under the provisions of Section 10 of this article.

When considering the above alternatives, the [superintendent] shall give consideration to the economic impact of each alternative on the discharger. If the [superintendent] permits the pretreatment or equalization of waste flows, the design and installation of the plants and equipment shall be subject to the review and approval of the [superintendent].

SEC. 6. Grease, oil, and sand interceptors shall be provided when, in the opinion of the [superintendent], they are necessary for the proper handling of liquid wastes containing floatable grease in excessive amounts, as specified in Section 4(*c*), or any flammable wastes, sand, or other harmful ingredients; except that such interceptors shall not be required for private living quarters or dwelling units. All interceptors shall be of a type and capacity approved by the [superintendent], and shall be located as to be readily and easily accessible for cleaning and inspection. In the maintaining of these interceptors the owner(s) shall be responsible for the proper removal and disposal by appropriate means of the captured material and shall maintain records of the dates, and means of disposal which are subject to review by the [superintendent]. Any removal and hauling of the collected materials not performed by owner(s)' personnel must be performed by currently licensed waste disposal firms.

SEC. 7. Where pretreatment or flowequalizing facilities are provided or required for any waters or wastes, they shall be maintained continuously in satisfactory and effective operation by the owner(s) at his expense.

SEC. 8. When required by the [superintendent], the owner of any property serviced by a building sewer carrying industrial wastes shall install a suitable structure together with such necessary meters and other appurtenances in the building sewer to facilitate observation, sampling, and measurement of the wastes. Such structure, when required, shall be accessibly and safely located and shall be constructed in accordance with plans approved by the [superintendent]. The structure shall be installed by the owner at his expense and shall be maintained by him so as to be safe and accessible at all times.

SEC. 9. The [superintendent] may require a user of sewer services to provide information needed to determine compliance with this ordinance. These requirements may include:

(1) Wastewaters discharge peak rate and volume over a specified time period.
(2) Chemical analyses of wastewaters.
(3) Information on raw materials, processes, and products affecting wastewater volume and quality.
(4) Quantity and disposition of specific liquid, sludge, oil, solvent, or other materials important to sewer use control.
(5) A plot plan of sewers of the user's property showing sewer and pretreatment facility location.
(6) Details of watewater pretreatment facilities.
(7) Details of systems to prevent and control the losses of materials through spills to the municipal sewer.

SEC. 10. All measurements, tests, and analyses of the characteristics of waters and wastes to which reference is made in this ordinance shall be determined in accordance with the latest edition of *Standard Methods for the Examination of Water and Wastewater,* published by the American Public Health Association. Sampling methods, location, times, durations, and frequencies are to be determined on an individual basis subject to approval by the [superintendent].

SEC. 11. No statement contained in this article shall be construed as preventing any special agreement or arrangement between the [city] and any industrial concern whereby an industrial waste of unusual strength or character may be accepted by the [city] for treatment.

ARTICLE VI

SEC. 1. No person(s) shall maliciously, willfully, or negligently break, damage, destroy, uncover, deface, or tamper with any structure, appurtenance or equipment which is a part of the wastewater facilities. Any person(s) violating this provision shall be subject to immediate arrest under charge of disorderly conduct.

ARTICLE VII

Powers and Authority of Inspectors

SEC. 1. The [superintendent] and other duly authorized employees of the [city] bearing proper credentials and identification shall be permitted to enter all properties for the purposes of inspection, observation, measurement, sampling, and testing pertinent to discahrge to the community system in accordance with the provisions of this ordinance.

SEC. 2. The [superintendent] or other duly authorized employees are authorized to obtain information concerning industrial processes which have a direct bearing on the kind and source of discharge to the wastewater collection system. The industry may withhold information considered confidential. The industry must establish that the revelation to the public of the information in question might result in an advantage to competitors.

SEC. 3. While performing the necessary work on private properties referred to in Article XII, Section 1, above, the [superintendent] or duly authorized employees of the [city] shall observe all safety rules applicable to the premises established by the company, and the company shall be held harmless for injury or death to the [city] employees, and the [city] shall indemnify the company against loss or damage to its property by [city] employees and against liability claims and demands for personal injury or property damage asserted against the company and growing out of the gauging and sampling operation, except as such may be caused by negligence or failure of the company to maintain safe conditions as require in Article V, Section 8.

SEC. 4. The [superintendent] and other duly authorized employees of the [city] bearing proper credentials and identification shall be permitted to enter all private properties through which the [city] holds a duly negotiated easement for the purposes of, but not limited to, inspection, observation, measurement, sampling, repair, and maintenance of any portion of the wastewater facilities lying within said easement. All entry and subsequent work, if any, on said easement, shall be done in full accordance with the terms of the duly negotiated easement pertaining to the private property involved.

Suggested Permit Application Forms

Sample application forms for the three classes of permits for sewer service are given below.

Private Waste Disposal Application

To the [city] of ....:
    The undersigned, being the .... of the
(owner, owner's agent)

property located at ....
(number)   (street)

does hereby request a permit to install sanitary sewage disposal facilities to serve
the .... at the location.
(residence, commercial building, etc.)

1. The proposed facilities include: ....
....
to be constructed in complete accordance with the plans and specifications attached hereunto as Exhibit "A".

2. The area of the property is .... square feet (or square meters).

3. The name and address of the person or firm who will perform the work is:
....

4. The maximum number of persons to be served by the proposed facilities is
....

5. The locations and nature of all sources of private or public water supply within one hundred (100) feet (30.5 meters) of any boundary of said property are shown on the plat attached hereunto as Exhibit "B."

In consideration of the granting of this permit, the undersigned agrees:

1. To furnish any additional information relating to the proposed work that shall be requested by the [superintendent].

2. To accept and abide by all provisions of Ordinance No. ....
of the [city] of ...., and of all other pertinent ordinances or regulations that may be adopted in the future.

3. To operate and maintain the wastewater disposal facilities covered by this application in a sanitary manner at all times, in compliance with all requirements of the [health officer], and at no expense to the [city].

4. To notify the [superintendent] at least twenty-four (24) hours prior to commencement of the work proposed, and again at least twenty-four (24) hours prior to the covering of any underground portions of the installation.

Date: ....   Signed ....
                                              (applicant)

                                      ....
                                       (address of applicant)

$............... inspection fee paid.   ....
Application approved and permit   (certification by [city] treasurer)
issued:

Date ....   Signed ....
                                              (superintendent)

Residential or Commercial Building Sewer Application

To the [city] of ....:

The undersigned, being the .... of the
                                      (owner, owner's agent)
property located at ...., does
                          (number)   (street)
hereby request a permit to install and connect a building sewer to serve the
.... at said location.
        (residence, commercial building, etc.)

1. The following indicated fixtures will be connected to the proposed building sewer:

Number	*Fixture*	*Number*	*Fixture*
....	Kitchen sinks	....	Water closets
....	Lavatories	....	Bath tubs
....	Laundry tubs	....	Showers
....	Urinals	....	Garbage grinders

Specify other fixtures ....

2. The maximum number of persons who will use the above fixtures is ....

3. The name and address of the person or firm who will perform the proposed work is ....

4. Plans and specifications for the proposed building sewer are attached hereunto as Exhibit "A."

In consideration of the granting of this permit, the undersigned agrees:

1. To accept and abide by all provisions of Ordinance No. ....
of the [city] of ...., and of all other pertinent ordinances or regulations that may be adopted in the future.

2. To maintain the building sewer at no expense to the [city].

3. To notify the [superintendent] when the building sewer is ready for inspection and connection to the public sewer, but before any portion of the work is covered.

Date: ....   Signed ....

(applicant)

....

$................ inspection fee paid.   (address of applicant)
Application   approved   and   permit   ....
issued: (certification by [city] treasurer)
Date: ....   Signed ....

(superintendent)

Industrial Sewer Connection Application

To the [city] of ....:

The undersigned being the .... of the
(owner, lessee, tenant, etc.)

property located at ....
....

does hereby request a permit to .... an industrial
(install, use)

sewer connection serving the ...., which company
(name of company)

is engaged in ....
....

at said location.

1. A plan of the property showing accurately all sewers and drains now existing is attached hereunto as Exhibit "A."

2. Plans and specifications covering any work proposed to be performed under this permit is attached hereunto as Exhibit "B."

3. A complete schedule of all process waters and industrial wastes produced or expected to be produced at said property, including a description of the character of each waste, the daily volume and maximum rates of discharge, and representative analyses, is attached hereunto as Exhibit "C."

4. The name and address of the person or firm who will perform the work covered by this permit is . . . .
. . . .

In consideration of the granting of this permit the undersigned agrees:

1. To furnish any additional information relating to the installation or use of the industrial sewer for which this permit is sought as may be requested by the [superintendent].

2. To accept and abide by all provisions of Ordinance No. . . . .
of the [city] of . . . . , and of all other pertinent ordinances or regulations that may be adopted in the future.

3. To operate and maintain any waste pretreatment facilities, as may be required as a condition of the acceptance into the public sewer of the industrial wastes involved, in an efficient manner at all times, and at no expense to the [city].

4. To cooperate at all times with the [superintendent] and his representatives in their inspecting, sampling, and study of the industrial wastes, and any facilities provided for pretreatment.

5. To notify the [superintendent] immediately in the event of any accident, negligence, or other occurrence that occasions discharge to the public sewers of any wastes or process waters not covered by this permit.

Date . . . .   Signed . . . .

                                      (applicant)

               . . . .

$. . . . . inspection fee paid.   (address of applicant)

Application approved and permit   . . . .
granted: (certification by [city] treasurer)

Date: . . . .   Signed . . . .

                                   (superintendent)

FEDERAL PRETREATMENT GUIDELINES: RECOMMENDED ORDINANCE FOR INDUSTRIAL USE OF PUBLICLY OWNED SEWERAGE FACILITIES

Ordinance No. ________

An ordinance establishing rules and regulations for the discharge of wastewaters into the wastewater treatment system of the City of ________________________________.

Whereas, the Federal Water Pollution Control Act Amendments of 1972, P.L. 92-500 (hereinafter referred to as the "Act") have resulted in an unprecedented program of cleaning up our Nation's waters;

Whereas, this City has already made and will continue to make a substantial financial investment in its wastewater treatment system to achieve the goals of the Act; and,

Whereas, this City seeks to provide for the use of its wastewater treatment system by industries served by it without damage to the physical facilities, without impairment of their normal function of collecting, treating and discharging domestic wastewater, and without the discharge by this City's wastewater treatment system of pollutants which would violate the discharge allowed under its National Pollutant Discharge Elimination System (NPDES) permit and the applicable rules of all governmental authorities with jurisdiction over such discharges.

Now, therefore, be it ordained and enacted by the City Council of the City of ________________, County of ________________, State of ________________, as follows:

SECTION 1

Definitions

Unless the context specifically indicates otherwise, the following terms, as used in this Ordinance, shall have the meanings hereinafter designated:

(*a*) "**BIOCHEMICAL OXYGEN DEMAND**" (BOD) means the quantity of oxygen utilized in the biochemical oxidation of organic matter under standard laboratory procedure in five (5) days at 20° C, expressed in terms of weight and concentration (milligrams per liter).

(*b*) "**COOLING WATER**" means the water discharged from any use such as air conditioning, cooling or refrigeration, during which the only pollutant added to the water is heat.

(*c*) "**COMPATIBLE POLLUTANT**" means BOD, suspended solids, *p*H, and fecal coliform bacteria, and such additional pollutants as are now or may be in the future specified and controlled in this City's NPDES permit for its wastewater treatment works where said works have been designed and used to reduce or remove such pollutants.

(*d*) "**DIRECTOR**"[/"Superintendent"] means the [director/superintendent of wastewater treatment system/of water pollution control/of public works] of this City or his duly appointed deputy, agent or representative.

(*e*) "**DOMESTIC WASTES**" means liquid wastes (i) from the noncommercial preparation, cooking, and handling of food or (ii) containing human excrement and similar matter from the sanitary conveniences of dwellings, commercial buildings, industrial facilities, and institutions.

(*f*) "**GARBAGE**" means solid wastes from domestic and commercial preparation, cooking and dispensing of food, and from the handling, storage, and sale of food.

(*g*) "**INCOMPATIBLE POLLUTANT**" means any pollutant which is not a "compatible pollutant" as defined in this section.

(*h*) "**INDUSTRIAL WASTEWATER**" means the liquid wastes resulting from the processes employed in industrial, manufacturing, trade, or business establishments, as distinct from domestic wastes.

(*i*) "**NATIONAL POLLUTANT DISCHARGE ELIMINATION SYSTEM**" (NPDES) means the program for issuing, conditioning, and denying permits for the discharge of pollutants from point sources into the navigable waters, the contiguous zone and the oceans pursuant to section 402 of the Act.

(*j*) "**PERSON**" means any individual, firm, company, partnership, corporation, association, group, or society, and includes the State of ________________________, and agencies, districts, commissions and political subdivisions created by or pursuant to State law.

(*k*) "**pH**" means the logarithm of the reciprocal of the concentration of hydrogen ions in grams per liter of solution.

(*l*) "**PRETREATMENT**" means application of physical, chemical, and biological processes to reduce the amount of pollutants in or alter the nature of the pollutant properties in a wastewater prior to discharging such wastewater into the publicly owned wastewater treatment system.

(*m*) "**PRETREATMENT STANDARDS**" means all applicable Federal rules and regulations implementing section 307 of the Act, as well as any nonconflicting State or local standards. In cases of conflicting standards or regulations, the more stringent thereof shall be applied.

(*n*) "**SIGNIFICANT INDUSTRIAL USER**" means any industrial user of the City's wastewater treatment system whose flow exceeds (i) [50,000] gallons per day, or (ii) [five (5)] percent of the daily capacity of the treatment system.

(*o*) "**STORM WATER**" means any flow occurring during or immediately following any form of natural precipitation and resulting therefrom.

(*p*) "**SUSPENDED SOLIDS**" means the total suspended matter that floats on the surface of, or is suspended in, water, wastewater, or other liquids, and which is removable by laboratory filtering.

(*q*) "**UNPOLLUTED WATER**" is water not containing any pollutants limited or prohibited by the effluent standards in effect, or water whose discharge will not cause any violation of receiving water quality standards.

(*r*) "**USER**" means any person who discharges or causes or permits the discharge of wastewater into the City's wastewater treatment system.

(*s*) "**USER CLASSIFICATION**" means a classification of user based on the 1972 (or subsequent) edition of the Standard Industrial Classification (SIC) Manual prepared by the Office of Management and Budget.

(*t*) "**WASTEWATER**" means the liquid and water-carried industrial or domestic wastes from dwellings, commercial buildings, industrial facilities, and institutions, together with any groundwater, surface water, and storm water that may be present, whether treated or untreated, which is discharged into or permitted to enter the City's treatment works.

(*u*) "**WASTEWATER TREATMENT SYSTEM**" (system) means any devices, facilities, structures, equipment or works owned or used by the City for the purpose of the transmission, storage, treatment, recycling, and reclamation of industrial and domestic wastes, or necessary to recycle or reuse water at the most economical cost over the estimated life of the system, including intercepting sewers, outfall sewers, sewage collection systems, pumping, power, and other equipment, and their appurtenances; extensions, improvements, remodeling, additions, and alterations thereof; elements essential to provide a reliable recycled supply such as standby treatment units and clear well facilities; and any works, including site acquisition of the land that will be an integral part of the treatment process or is used for ultimate disposal of residues resulting from such treatment.

(*v*) Terms not otherwise defined herein shall be as adopted in the latest edition of *Standard Methods for the Examination of Water and Wastewater,* published by the American Public Health Association, the American Water Works Association, and the Water Pollution Control Federation.

SECTION 2

Prohibitions and Limitations on Wastewater Discharges

(*a*) **PROHIBITIONS ON WASTEWATER DISCHARGES.** No person shall discharge or deposit or cause or allow to be discharged or deposited into the wastewater treatment system any wastewater which contains the following:

(1) *Oils and grease.* (A) Oil and grease concentrations or amounts from industrial facilities violating Federal pretreatment standards. (B) Wastewater from industrial facilities containing floatable fats, wax, grease or oils. [Optional: (C) Wax, grease or oil concentration of mineral origin of more than (   ) mg/l whether emulsified or not, or containing substances which may solidify or become viscous at temperatures between 32° and 150°F (0° and 65°C) at the point of discharge into the system.] [Optional: (D) Total fat, wax, grease or oil concentration of more than (   )mg/l, whether emulsified or not, or containing substances which may solidify or become viscous at temperatures between 32° and 150°F (0° and 65°C) at the point of discharge into the system.]

(2) *Explosive mixtures.* Liquids, solids or gases which by reason of their nature or quantity are, or may be, sufficient either alone or by interaction with other substances to cause fire or explosion or be injurious in any other way to the sewerage facilities or to the operation of the system. At no time shall two successive readings on an explosion hazard meter, at the point of discharge into the sewer system, be more than five percent (5%) nor any single reading over ten percent (10%) of the Lower Explosive Limit (L.E.L.) of the meter. Prohibited materials include, but are not limited to, gasoline, kerosene, naphtha, benzene, toluene, xylene, ethers, alcohols, ketones, aldehydes, peroxides, chlorates, perchlorates, bromates, carbides, hydrides, and sulfides.

(3) *Noxious material.* Noxious or malodorous solids, liquids, or gases, which, either singly or by interaction with other wastes, are capable of creating a public nuisance or hazard to life, or are or may be sufficient to prevent entry into a sewer for its maintenance and repair.

(4) *Improperly shredded garbage.* Garbage that has not been ground or comminuted to such a degree that all particles will be carried freely in suspension under flow conditions normally prevailing in the public sewers, with no particle greater than one-half (½) inch in any dimension.

(5) *Radioactive wastes.* Radioactive wastes or isotopes of such half-life or concentra-

tion that they do not comply with regulations or orders issued by the appropriate authority having control over their use and which will or may cause damage or hazards to the sewerage facilities or personnel operating the system.

(6) *Solid or viscous wastes.* Solid or viscous wastes which will or may cause obstruction to the flow in a sewer, or otherwise interfere with the proper operation of the wastewater treatment system. Prohibited materials include, but are not limited to, grease, uncomminuted garbage, animal guts or tissues, paunch manure, bones, hair, hides or fleshings, entrails, whole blood, feathers, ashes, cinders, sand, spent lime, stone or marble dust, metal, glass, straw, shavings, grass clippings, rags, spent grains, spent hops, waste paper, wood, plastic, tar, asphalt residues, residues from refining or processing of fuel or lubricating oil, and similar substances.

(7) *Excessive discharge rate.* Wastewaters at a flow rate or containing such concentrations or quantities of pollutants that exceeds for any time period longer than fifteen (15) minutes more than five (5) times the average twenty-four (24) hour concentration, quantities or flow during normal operation and that would cause a treatment process upset and subsequent loss of treatment efficiency.

(8) *Toxic substances.* Any toxic substances in amounts exceeding standards promulgated by the Administrator of the United States Environmental Protection Agency pursuant to section 307(a) of the Act, and chemical elements or compounds, phenols or other taste- or odor-producing substances, or any other substances which are not susceptible to treatment or which may interfere with the biological processes or efficiency of the treatment system, or that will pass through the system.

(9) *Unpolluted waters.* Any unpolluted water including, but not limited to, water from cooling systems or of stormwater origin, which will increase the hydraulic load on the treatment system.

(10) *Discolored material.* Wastes with objectionable color not removable by the treatment process.

(11) *Corrosive wastes.* Any waste which will cause corrosion or deterioration of the treatment system. All wastes discharged to the public sewer system must have a pH value in the range of [6] to [9] standard units. Prohibited materials include, but are not limited to, acids, sulfides, concentrated chloride and fluoride compounds, and substances which will react with water to form acidic products.

(*b*) **LIMITATIONS ON WASTEWATER DISCHARGES.** [Use either Option A or Option B.]

[Option A   General Limitations]

No person shall discharge or convey, or permit or allow to be discharged or conveyed, to a public sewer any wastewater containing pollutants of such character or quantity that will:

(1) Not be susceptible to treatment or interfere with the process or efficiency of the treatment system.

(2) Constitute a hazard to human or animal life, or to the stream or water course receiving the treatment plant effluent.

(3) Violate pretreatment standards.

(4) Cause the treatment plant to violate its NPDES permit or applicable receiving water standards.

[*Option B   Specific Limitations*]

Restrictions on the maximum concentration (mg/l) or mass limitations (kg/kkg) allowable in wastewater discharge to the wastewater treatment system can be found in sources such as Federal Guidelines or State and Local Pretreatment Programs (Volume I, Section C) for the following pollutants: arsenic, barium, boron, cadmium, chromium (total), chromium (trivalent), chromium (hexavalent), chlorinated hydrocarbons, copper, cyanide, iron, lead, manganese, mercury, nickel, phenolic compounds, phosphorus, selenium, silver, surfactants, zinc. There are also restrictions on pH and temperature (not over 150° F, except where higher temperatures are permitted by law) of wastewater discharges. Dilution of any wastewater discharge for the purpose of satisfying these requirements shall be considered a violation of this Ordinance.

(*c*) SPECIAL AGREEMENTS. Nothing in this section shall be construed as preventing any special agreement or arrangement between the City and any user of the wastewater treatment system whereby wastewater of unusual strength or character is accepted into the system and specially treated subject to any payments or user charges as may be applicable.

SECTION 3

Control of Prohibited Wastes

(*a*) REGULATORY ACTIONS. If wastewaters containing any substance described in Section 2 of this ordinance are discharged or proposed to be discharged into the sewer system of the City or to any sewer system tributory thereto, the Director and [Corporation Counsel/City Attorney] may take any action necessary to:
(1) Prohibit the discharge of such wastewater.
(2) Require a discharger to demonstrate that in-plant modifications will reduce or eliminate the discharge of such substances in conformity with this Ordinance.
(3) Require pretreatment, including storage facilities, or flow equalization necessary to reduce or eliminate the objectionable characteristics or substances so that the discharge will not violate these rules and regulations.
(4) Require the person making, causing or allowing the discharge to pay any additional cost or expense incurred by the City for handling and treating excess loads imposed on the treatment system.
(5) Take such other remedial action as may be deemed to be desirable or necessary to achieve the purpose of this Ordinance.
(*b*) SUBMISSION OF PLANS. Where pretreatment or equalization of wastewater flows prior to discharge into any part of the wastewater treatment system is required, plans, specifications and other pertinent data or information relating to such pretreatment or flow-control facilities shall first be submitted to the Director for review and approval. Such approval shall not exempt the discharge or such facilities from compliance with any applicable code, ordinance, rule, regulation or order of any governmental authority. Any subsequent alterations or additions to such pretreatment or flow-control facilities shall not be made without due notice to and prior approval of the Director.

(*c*) **PRETREATMENT FACILITIES OPERATIONS.** If pretreatment or control of waste flows is required, such facilities shall be maintained in good working order and operated as efficiently as possible by the owner or operator at his own cost and expense, subject to the requirements of these rules and regulations and all other applicable codes, ordinances, and laws.

(*d*) **ADMISSION TO PROPERTY.** Whenever it shall be necessary for the purposes of these rules and regulations, the Director, upon the presentation of credentials, may enter upon any property or premises at reasonable times for the purpose of (1) copying any records required to be kept under the provisions of this Ordinance, (2) inspecting any monitoring equipment or method, and (3) sampling any discharge of wastewater to the treatment works. The Director may enter upon the property at any hour under emergency circumstances.

(*e*) **PROTECTION FROM ACCIDENTAL DISCHARGE.** Each industrial user shall provide protection from accidental discharge of prohibited materials or other wastes regulated by this Ordinance. Facilities to prevent accidental discharge of prohibited materials shall be provided and maintained at the owner or operator's own cost and expense. Detailed plans showing facilities and operating procedures to provide this protection shall be submitted to the Director for review, and shall be approved by him before construction of the facility. Review and approval of such plans and operating procedures shall not relieve the industrial user from the responsibility to modify his facility as necessary to meet the requirements of this Ordinance.

(*f*) **REPORTING OF ACCIDENTAL DISCHARGE.** If, for any reason, a facility does not comply with or will be unable to comply with any prohibition or limitations in this Ordinance, the facility responsible for such discharge shall immediately notify the Director so that corrective action may be taken to protect the treatment system. In addition, a written report addressed to the Director detailing the date, time and cause of the accidental discharge, the quantity and characteristics of the discharge, and corrective action taken to prevent future discharges, shall be filed by the responsible industrial facility within five (5) days of the occurrence of the noncomplying discharge.

SECTION 4

Industrial Wastewater Monitoring and Reporting

(*a*) **DISCHARGE REPORTS.**
(1) Every significant industrial user shall file a periodic Discharge Report at such intervals as are designated by the Director. The Director may require any other industrial users discharging or proposing to discharge into the treatment system to file such periodic reports.
(2) The discharge report shall include, but, in the discretion of the Director, shall not be limited to, nature of process, volume, rates of flow, mass emission rate, production quantities, hours of operation, concentrations of controlled pollutants, or other information which relates to the generation of waste. Such reports may also include the chemical constituents and quantity of liquid materials stored on site even though they are not normally discharged. In addition to

discharge reports, the Director may require information in the form of [Industrial Discharge Permit Applications and (optional)] self-monitoring reports.

(*b*) **RECORDS AND MONITORING.**

(1) All industrial users who discharge or propose to discharge wastewaters to the wastewater treatment system shall maintain such records of production and related factors, effluent flows, and pollutant amounts or concentrations as are necessary to demonstrate compliance with the requirements of this Ordinance and any applicable State or Federal pretreatment standards or requirements.

(2) Such records shall be made available upon request by the Director. All such records relating to compliance with pretreatment standards shall be made available to officials of the U.S. Environmental Protection Agency upon demand. A summary of such data indicating the industrial user's compliance with this Ordinance shall be prepared [quarterly] (optional) and submitted to the Director.

(3) The owner or operator of any premises or facility discharging industrial wastes into the system shall install at his own cost and expense suitable monitoring equipment to facilitate the accurate observation, sampling, and measurement of wastes. Such equipment shall be maintained in proper working order and kept safe and accessible at all times.

(4) The monitoring equipment shall be located and maintained on the industrial user's premises outside of the building. When such a location would be impractical or cause undue hardship on the user, the Director may allow such facility to be constructed in the public street or sidewalk area, with the approval of the public agency having jurisdiction over such street or sidewalk, and located so that it will not be obstructed by public utilities, landscaping, or parked vehicles.

(5) When more than one user can discharge into a common sewer, the Director may require installation of separate monitoring equipment for each user. When there is a significant difference in wastewater constituents and characteristics produced by different operations of a single user, the Director may require that separate monitoring facilities be installed for each separate discharge.

(6) Whether constructed on public or private property, the monitoring facilities shall be constructed in accordance with the Director's requirements and all applicable construction standards and specifications.

(*c*) **INSPECTION, SAMPLING AND ANALYSIS.**

(1) *Compliance determination.* Compliance determinations with respect to Section 2 prohibitions and limitations may be made on the basis of either instantaneous grab samples or composite samples of wastewater. Composite samples may be taken over a 24 hour period, or over a longer or shorter time span, as determined necessary by the Director to meet the needs of specific circumstances.

(2) *Analysis of industrial wastewaters.* Laboratory analysis of industrial wastewater samples shall be performed in accordance with the current edition of *Standard Methods, Methods for Chemical Analysis of Water and Waste* published by the U.S. Environmental Protection Agency or the *Annual Book of Standards, Part 23, Water, Atmospheric Analysis,* published by the American Society for Testing and Materials. Analysis of those pollutants not covered by these publications shall be performed in accordance with procedures established by the [State Department of Environmental Health.]

(3) *Sampling frequency* [Optional]. Sampling of industrial wastewater for the purpose of compliance determination with respect to Section 2 prohibitions and limitations will be done at such intervals as the Director may designate. However, it is the intention of the Director to conduct compliance sampling or to cause such sampling to be conducted for all major contributing industries at least once in every [1 year] (optional) period.

SECTION 5

Industrial Discharge Permit System [Optional]

(*a*) **WASTEWATER DISCHARGE PERMITS REQUIRED.** All significant industrial users proposing to connect to or discharge into any part of the wastewater treatment system must first obtain a discharge permit. All existing significant industrial users connected to or discharging to any part of the City system must obtain a wastewater discharge permit within ninety (90) (optional) days from and after the effective date of this Ordinance.

(*b*) **PERMIT APPLICATION.** Users seeking a wastewater discharge permit shall complete and file with the Director an application on the form prescribed by the Director, and accompanied by the applicable fee. In support of this application, the user shall submit the following information:
(1) Name, address, and SIC number of applicant.
(2) Volume of wastewater to be discharged.
(3) Wastewater constituents and characteristics including, but not limited to, those set forth in Section 2 of this Ordinance as determined by a reliable analytical laboratory.
(4) Time and duration of discharge.
(5) Average and [30] (optional) minute peak wastewater flow rates, including daily, monthly and seasonal variations, if any.
(6) Site plans, floor plans, mechanical and plumbing plans and details to show all sewers and appurtenances by size, location and elevation.
(7) Description of activities, facilities and plant processes on the premises including all materials and types of materials which are, or could be, discharged.
(8) Each product produced by type, amount, and rate of production.
(9) Number and type of employees, and hours of work.
(10) Any other information as may be deemed by the Director to be necessary to evaluate the permit application.
The Director will evaluate the data furnished by the user and may require additional information. After evaluation and acceptance of the data furnished, the Director may issue a wastewater discharge permit subject to terms and conditions provided herein.

(*c*) **PERMIT CONDITIONS.** Wastewater discharge permits shall be expressly subject to all provisions of this Ordinance and all other regulations, user charges, and fees established by the City. The conditions of wastewater discharge permits shall be uniformly enforced in accordance with this Ordinance, and applicable State and Federal regulations. Permit conditions will include the following:
(1) The unit charge or schedule of user charges and fees for the wastewater to be discharged to the system.

(2) The average and maximum wastewater constituents and characteristics.

(3) Limits on rate and time of discharge or requirements for flow regulations and equalization.

(4) Requirements for installation of inspection and sampling facilities, and specifications for monitoring programs.

(5) Requirements for maintaining and submitting technical reports and plant records relating to wastewater discharges.

(6) Daily average and daily maximum discharge rates, or other appropriate conditions when pollutants subject to limitations and prohibitions are proposed or present in the user's wastewater discharge.

(7) Compliance schedules.

(8) Other conditions to ensure compliance with this Ordinance.

(*d*) **DURATION OF PERMITS.** Permits shall be issued for a specified time period, not to exceed [five] (optional) years. A permit may be issued for a period of less than [one] (optional) year, or may be stated to expire on a specific date. If the user is not notified by the Director [30] (optional) days prior to the expiration of the permit, the permit shall automatically be extended for [   ] months. The terms and conditions of the permit may be subject to modification and change by the [responsible official] during the life of the permit, as limitations or requirements as identified in Section 2 are modified and changed. The user shall be informed of any proposed changes in his permit at least [30] (optional) days prior to the effective date of change. Any changes or new conditions in the permit shall include a reasonable time schedule for compliance.

(*e*) **TRANSFER OF A PERMIT.** Wastewater discharge permits are issued to a specific user for a specific operation. A wastewater discharge permit shall not be reassigned or transferred or sold to a new owner, new user, different premises, or a new or changed operation.

(*f*) **REVOCATION OF PERMIT.** Any user who violates the following conditions of his permit or of this Ordinance, or of applicable State and Federal regulations, is subject to having his permit revoked. Violations subjecting a user to possible revocation of his permit include, but are not limited to, the following:

(1) Failure of a user to accurately report the wastewater constituents and characteristics of his discharge;

(2) Failure of the user to report significant changes in operations, or wastewater constituents and characteristics;

(3) Refusal of reasonable access to the user's premises for the purpose of inspection or monitoring; or,

(4) Violation of conditions of the permit.

SECTION 6

Enforcement Procedures

(*a*) **NOTIFICATION OF VIOLATION.** Whenever the Director finds that any person has violated or is violating this Ordinance, or any prohibition, limitation, or requirement contained herein, he may serve upon such person a written notice stating the nature of the violation and providing a reasonable time, not to exceed thirty (30) days, for the satisfactory correction thereof.

(*b*) **Show Cause Hearing.**

(1) If the violation is not corrected by timely compliance, the Director may order any person who causes or allows an unauthorized discharge to show cause before the [hearing authority] why service should not be terminated. A notice shall be served on the offending party, specifying the time and place of a hearing to be held by the [hearing authority] regarding the violation, and directing the offending party to show cause before [said authority] why an order should not be made directing the termination of service. The notice of the hearing shall be served personally or by registered or certified mail (return receipt requested) at least [ten] days before the hearing. Service may be made on any agent or officer of a corporation.

(2) The [hearing authority] may itself conduct the hearing and take the evidence, or may designate any of its members or any officer or employee of the [assigned department] to:

(A) Issue in the name of the [hearing authority] notices of hearings requesting the attendance and testimony of witnesses and the production of evidence relevant to any matter involved in any such hearings.

(B) Take the evidence.

(C) Transmit a report of the evidence and hearing, including transcripts and other evidence, together with recommendations to the [hearing authority] for action thereon.

(3) At any public hearing, testimony taken before the hearing authority or any person designated by it, must be under oath and recorded stenographically. The transcript, so recorded, will be made available to any member of the public or any part to the hearing upon payment of the usual charges.

(4) After the [hearing authority] has reviewed the evidence, it may issue an order to the party responsible for the discharge directing that, following a specified time period, the sewer service be discontinued unless adequate treatment facilities, devices or other related appurtenances shall have been installed or existing treatment facilities, devices, or other related appurtenances are properly operated, and such further orders and directives as are necessary and appropriate.

(*c*) **Legal Action.** Any discharge in violation of the substantive provisions of this Ordinance or an Order of the [hearing authority] shall be considered a public nuisance. If any person discharges sewage, industrial wastes or other wastes into the City treatment system contrary to the substantive provisions of this Ordinance or any Order of the [hearing authority], the [Corporation Counsel/City Attorney] shall commence an action for appropriate legal and/or equitable relief in the [Circuit] Court of this County.

Section 7

Penalty; Costs

Any person who is found to have violated an Order of the [hearing authority] or who willfully or negligently failed to comply with any provision of this Ordinance, and the orders, rules, and regulations issued hereunder, shall be fined not less than [One Hundred Dollars] (optional) nor more than [One Thousand Dollars] (optional) for each offense. Each day on which a violation shall occur or continue shall be deemed a separate and distinct offense. In addition to the penalties provided herein, the City may recover reasonable attorneys' fees, court costs, court reporters' fees and other expenses of litigation by appropriate suit at law against the person found to have violated this Ordinance or the orders, rules and regulations issued hereunder.

Section 8

Savings Clause

If any provision, paragraph, word, section or article of this Ordinance is invalidated by any court of competent jurisdiction, the remaining provisions, paragraphs, words, sections, and articles shall not be affected and shall continue in full force and effect.

Section 9

Conflict

All ordinances and parts of ordinances inconsistent or conflicting with any part of this Ordinance are hereby repealed to the extent of such inconsistency or conflict.

Section 10

Effective Date

This Ordinance shall be in full force and effect [Option A] from and after its passage, approval and publication, as provided by law. [Option B] on the _____ day of _____, 19_____.

Introduced the _____ day of _____, 19_____.

First reading: _________, 19_____.
Second reading: _________, 19_____.

Passed this _____ day of _____, 19_____.

Ayes:
Nays:
Absent:
Not Voting:

Approved by me this _____ day of _____, 19_____.

____
                                          Mayor
Attest: _____________________ (Seal) City Clerk
Published the _____ day of _____, 19_____.

4

Impact of Industrial Wastes on Municipal Wastewater Treatment

BACKGROUND

In this chapter, the effect of industrial pollutants on POTW operations will be covered. In general, three types of impact may be expected. Pollutants associated with industrial wastewaters may interfere with POTW operations, reducing the efficiency of the system. Secondly, industrial pollutants, which the POTW may not have been designed to treat, may pass through a treatment plant causing added pollutant load on the receiving streams. Finally, incidental removal may result in toxic pollutant contamination of sludges. Passthrough, incidental removal, and POTW interference are addressed in this chapter.

It should be noted that the majority of research information on the inhibition of biological treatment process is for laboratory experiments, not full-scale operations. In general, although many toxic materials have been found to inhibit biological processes, the levels at which these pollutants are found in POTW influents is commonly lower than experimentally determined inhibitory levels. Nevertheless, at high enough concentrations treatment plant upset and significant pollutant passthrough have been reported. Similarly, much of the available data on toxic pollutant removed in POTWs is based on self-monitoring data. As a result, certain data inconsistencies are unavoidable because of the variable sampling and analytical methods used.

INTERFERENCE AND INHIBITION OF BIOLOGICAL TREATMENT PROCESSES

Introduction

Interference with the operation of a POTW system can be caused by a wide variety of chemical, biological and physical phenomena. In a broad sense, interference consists not only of materials which inhibit biological sewage treatment processes, but also substances which cause problems in sewage collection systems, sludge disposal or utilization methods, water reuse, land application of wastewater, or other operations. Collection system problems include fire and explosion hazards, corrosion, excessive heat, and solid or viscous wastes which cause plugging of sewers. Interference with sludge disposal or utilization and reuse of wastewaters consists primarily of incompatible or toxic pollutants which are concentrated in sludge, limiting reuse options or disposal alternatives.

Mechanism for Biological Treatment Process Upset

There are four specific mechanisms through which biological treatment processes may be inhibited or upset by industrial pollutants. Once a biomass becomes acclimated or adjusted to its surroundings, reduced treatment efficiency can occur through any rapid change in the characteristics of the raw wastewater entering the plant. These rapid changes in the chemical or physical environment are often termed *shock loads*. The common types of shock loading are summarized below.

Materials Deleterious at Trace Levels

The introduction of specifically deleterious materials such as metallic pollutants (copper, zinc, chromium, etc.) at trace levels has a well-defined impact on treatment plant operation. These materials interfere with the metabolic activities within the biomass cells. Because of the precise nature of this type of shock load, it is possible to set down in advance the maximum acceptable concentration for such pollutants.

Qualitative Shock Loads

It has been reported that if a treatment system has become acclimated to a pollutant, the introduction of a new type of organic load can block the assimilation of the waste that the system was treating (1). This type of shock load is termed *qualitative* (2), and may be caused by the introduction to a system of a new industrial waste which would normally be biologically treatable.

Quantitative Shock Loads

Quantitative shock loads may be characterized as any sudden change in the BOD loading that a plant receives. Since the type of BOD loading is not changed, as in the case of qualitative shocks, the result is an inability of the POTW to treat all of the increased amount of organic matter entering the plant. Quantitative shock loads are often called *excessive discharge* to the POTW, and generally refer to the introduction of unusually large amounts of BOD to the system. Other oxygen demand parameters, such as COD or TOC, as well as suspended solids may be used to describe quantitative shock loads, but BOD is the most common measure of organic loading and therefore its use offers the most general description of the phenomenon.

Hydraulic Shock Loads

Hydraulic shock loads are generally characterized by a rapid decrease in the concentration of the waste or organic loading of the system (2). Such a decrease in loading may result from the sudden introduction of relatively clean cooling water or stormwater into the system, as is common with combined sewers. It should be noted that the initial portion of runoff from rainfall (the "first flush") in a combined sewer may contribute a large quantity of organic matter, actually increasing the BOD loading as the flow simultaneously increases. Hydraulic shock loads often cause operating difficulties in POTWs, resulting from the washout of solids from treatment plant unit processes.

Interference with POTW Operations

Certain classes of pollutants, which do not necessarily inhibit biological processes, can impair the overall operation of the POTW, including collection and pumping facilities. The federal pretreatment regulations list five classes of materials that may not be discharged into sewers. These five classes of prohibited pollutants include:

1. Pollutants that create a fire or explosion hazard;
2. Pollutants that will cause corrosive damage to the POTW, more specifically discharges with pH values below 5;
3. Solid or viscous pollutants in amounts that could cause obstruction in sewers or otherwise interfere with the operation of POTW:
4. Slug discharges or shock loads in terms of volume, strength, or oxygen demand, which the discharger knows to be of such magnitude as to cause treatment process upsets and subsequent loss of treatment efficiency;
5. Heat in amounts that will inhibit biological activity at POTW, specifically discharges that cause the temperature of the treatment plant influent to exceed 40°C or exceed 65°C at the point of contribution to the collection system.

Additionally, in general, the pretreatment regulations prohibit discharge of any pollutants that inhibit or interfere with POTW operations.

Although the discharges listed above may not inhibit biological processes, they can cause serious disruptions of treatment system operations. In the case of corrosive or explosive substances, severe damage to collection or treatment facilities may result from their introduction into the system. Unlike trace inhibitory materials that may be tolerable, or even beneficial at low levels, the prohibited materials included in this section should be closely regulated because of the severity of their effect on treatment and collection systems.

Corrosive Materials

Corrosion in relation to sewage systems can be defined as the phenomenon in which a pipe, conduit, or piece of equipment is gradually deteriorated by the fluid with which it comes in contact.

For sewage collection systems, one of the most prevalent types of corrosion is a deterioration of concrete pipe called *crown corrosion*. In this process, sulfates in wastewater are reduced under anaerobic conditions to sulfides, which hydrolyze to hydrogen sulfide. In addition, an industry such as an oil refinery or textile manufacturer may discharge sulfides directly to the sewer. If the concrete pipe is only partially filled, the H_2S formed may diffuse into the air above the fluid, where the bacteria *Thiobacillus* can convert it to sulfuric acid. Surfuric acid then reacts with the calcium oxide in the concrete, forming calcium sulfate or gypsum. This material is structurally unsound, and will eventually crumble.

Coating the interior of concrete pipes with an inert material can prevent crown corrosion, but an equally effective preventative measure is the exclusion of sulfides from the collection system. Chlorinating sulfide-carrying industrial wastewater is one way of achieving this goal. When the wastewater contains sulfate (e.g., in cases of noncarbonate hardness of the water supply), anaerobic conditions in the sewer should be avoided, since they provide the necessary atmosphere for the creation of H_2S. Anaerobic conditions, however, may be unavoidable in many instances; in those cases, proper sewer ventilation is necessary, not only to provide a measure of safety from potentially lethal hydrogen sulfide, but also to avoid the oxidation of H_2S to sulfuric acid. Although sulfide may be beneficial in terms of precipitating metals, its presence should be closely monitored to prevent crown corrosion.

The second type of corrosion is the dissolution of metallic pipes and structures. This commonly occurs when wastewater has a low pH, or contains some other oxidizing agent. Generally, waste mixtures should be pretreated to fall in the pH range of 6–9. Federal pretreatment regulations prohibit contributions of wastewaters with a pH of less than 5. Lower-pH (acidic) discharges will attack and disintegrate metal and concrete pipes and structures. Higher pH values are more tolerable, and enforcement of the high-pH restriction should be at the discretion of the POTW operator. In fact, since acidic discharges are by far the more prevalent problem, many municipalities tolerate alkaline discharges to offset the higher volume of acid wastewaters in the collection system.

High flouride levels, especially in acidic solutions, are a potential corrosion threat. High concentrations of chlorine, hypochlorites, and chlorides are all corrosive to metals. Many other products react with water to form acidic products which are detrimental to sewer systems. For example, acetic anhydride will hydrolize to form acetic acid, acetyl chloride will form acetic and hydrochloric acids, and ferric chloride will release hydrochloric acid upon hydrolysis. Sulfur dioxide and sulfurous acid are also common materials corrosive to sewers and sewage treatment plants.

Materials Which Cause Sewer Blockages

Discharges from commercial and industrial establishments must be controlled to prohibit materials which will clog sewers or treatment plant unit operation or form deposits that adversely affect hydraulic characteristics. Precipitates such as ferric hydrous oxide can interfere with the activated sludge system by inhibiting oxygen or food transfer between sludge particles and the surrounding liquids. Sulfates and carbonates can react with calcium salts to form a scale which can coat and ultimately block pipes. Fatty acids similarly react with calcium salts to form a curdy scum which can coat and block sewer lines.

Very high levels of suspended solids can cause blockages in sewer lines and overload primary settling tanks. Small fibers from textile industry operations may interfere with screens and filters by matting and blocking the passage of wastewaters. Grease accumulation can cause sewer line clogging both in the collection system and in the interconnecting sewers within the POTW. Collection systems with restaurant connections are especially prone to blockages, unless suitable grease collection facilities are provided at the contributing source.

Sewer clogging is most prevalent in small sewer lines, such as laterals, with 12-inch or smaller diameters. To cause a blockage problem, oil and grease, which is typically discharged in a liquid or emulsion form, must congeal or solidify. Generally, the temperatures encountered in sewers are high enough to prevent solidification of oil and grease of petroleum origin (hydrocarbon type). However, oil and grease of animal or vegetable origin can exhibit a broad range of congealing temperatures. Some oils and greases of animal or vegetable origin will congeal at the highest temperatures encountered in sewers, while others may remain fluid at the lowest temperatures. The probability that oil and grease will congeal in a sewer pipe depends on a number of factors, including the type of oil and grease involved, temperature, and the fluid velocity in the pipe. A sufficiently high velocity in the sewer can prevent clogging by keeping all congealed material in suspension, even if all other conditions are appropriate for solidification.

Explosive and Flammable Materials

The discharge of potentially explosive or flammable materials to sewer lines must be strictly controlled. A serious hazard can be created by hydrocarbon solvents, which float on the surface of water, and exert their full vapor pressure on the air space above. Such substances as gasoline, kerosene, naphtha, benzene, toluene, and xylene, therefore, are particularly hazardous. Ethers, alcohols, ketones, aldehydes, and organic peroxides similarly pose fire or explosion hazards.

Powerful oxidizing substances such as peroxides, chlorates, perchlorates, and bromates are potentially dangerous and should be restricted. Substances such as carbides, hydrides, and sulfides, which can liberate flammable or explosive gases, must also be carefully controlled.

Many of the above chemicals originate from facilities that manufacture or use organic chemicals. POTWs with such facilities among their contributors should give special attention to the control of explosive or flammable materials. The production of methane in sewers resulting from anaerobic conditions also presents a well-known potential explosion hazard. Because of the potential hazards caused by the presence of explosive materials in sewers, extreme care should be taken by anyone entering a sewer manhole.

Oil and Grease

The terms *oil* and *grease* cover a wide variety of substances that might be found in the influent to a POTW. Unlike other constituents, which usually have well-defined chemical forms, oil and grease characteristics are usually dependent on the method by which the material is analyzed. Oil and grease are thus best defined as those organic substances with similar solubilities in a particular extracting solvent.

The most commonly used solvents for oil and grease analysis are hexane, petroleum ether, and freon. These solvents will extract a broad spectrum of organic materials, including fatty acids, soaps, esters, fats, waxes, and various petroleum products (3). Freon is specifically recommended by EPA as the solvent for oil and grease extraction. Consequently, it is essential that all oil and grease measurements be completed utilizing the freon method to assure uniformity of results.

Oil and grease is a natural constituent of sewage. It has been reported that fecal material contains more than 25% grease (4). Additionally, domestic kitchen wastes contribute a large quantity of solvent-extractable materials. Commercial or industrial sources of oil and grease include slaughterhouses, food processors, and restaurants, as well as automobile service stations and petroleum refineries or storage depots. Although oil and grease from all of these sources may be extracted by the same solvent, and therefore considered a single constituent, the substances included in the analysis may or may not be biodegradable.

In general, the delineation between biodegradable oil and grease and the more refractory extractable substances corresponds to the distinction between oil and grease of animal and vegetable origin and oil and grease of petroleum origin. Oil and grease of animal and vegetable origin has been reported to be more biodegradable than oil and grease of petroleum origin (5). Nevertheless, many municipalities limit only total oil and grease (6). Traditionally, oil and grease concentration limits were focused on the prevention of sewer clogging and coating of pumping stations and treatment facilities, in which case the type of extractable material present was not important. However, distinguishing between oil and grease of petroleum and of animal or vegetable origin can yield information on both oil and grease treatability and source.

Another factor to be considered in the impact on POTW operation is the distinction between the two physical forms of oil and grease, namely floatable and dispersed or dissolved material. Many of the adverse effects experienced by POTWs, which are described below, result from the floatable portion of oil and grease present in wastewater. These problems can usually be avoided by employing gravity separation treatment techniques at the source. If such facilities are properly operated and maintained, floatable oil can be skimmed and thus kept out of the sewer system to a substantial degree.

In addition to partially passing through a biological treatment plant, petroleum oil at concentrations ranging between 50 and 100 mg/l has been reported to interfere with the aerobic processes in a POTW (7). It is believed that the principal interference is caused by attachment of oil, which has a density less than water, to the bacterial floc particles, which are to be gravity settled. The result is a slower settling rate, loss of solids by carryover out of the settling basin, and excessive release of BOD from the POTW to the environment (8). Additionally, in activated sludge units oil and grease may coat the biomass, interfering with oxygen transfer. As a consequence of this smothering action, a lower degree of treatment may be achieved (9). Oil and grease may also interfere with the operation of physical-chemical treatment facilities by coating the activated carbon, diminishing the adsorption of organic pollutants.

Oil and grease may also cause other problems in POTW operation. Actual operating problems have been reported in which oil and grease have clogged screens and interfered with skimming operations. Large quantities of oil and grease may block screens, scum drawoff systems, and sludge pumps, causing excessive loads on mechanical scraping and cleaning devices. Excessive oil and grease may also cause serious problems in pumping station operation by fouling float systems so that pumps fail to operate, and by blocking pump intakes. Additionally, oil and grease may foul electrodes used for monitoring volatile and explosive mixtures, causing a serious hazard. Since wetwells are seldom pumped dry, the accumulation of floating oil and grease can cause operational problems in this portion of the system as well (9).

Among the more troublesome operational problems caused by oil and grease are those associated with anaerobic digestor operation. Oil and grease can be responsible for foaming throughout the plant and especially in digestor vessels. If oil and grease reaches a covered digestor, a crust may form on the underside of the cover, causing serious maintenance problems, and reducing the available digestor volume. Scum layers in digestors may also interfere with effective mixing, temperature control, and gas separation (9).

IMPACT OF TOXIC POLLUTANTS ON MUNICIPAL SLUDGE QUALITY AND WASTEWATER RECYCLE

Interference with the biological unit processes of a POTW does not represent the only, or necessarily the most significant impact of industrial contributions. The sludges produced during the course of biological treatment will generally contain, in a concentrated form, many of the pollutants contributed to the POTW that may be considered inhibitory. Likewise, if pretreatment of industrial wastes is practiced, many of the undesirable industrial constituents undoubtedly will be concentrated in the industrial sludge. It should be noted that, although the municipal sludge generated by the POTW may not be considered a hazardous material, invariably the sludge generated by pretreatment of industrial wastes containing incompatible or toxic pollutants can be called hazardous. The responsibility for disposal of the industrial sludges does not lie with the POTW, but in general any assistance given to the pretreating industry will be valuable in terms of public relations and goodwill.

A detailed discussion of the impact of industrial wastewater or sludge handling, disposal and utilization is included in Chapter 5.

Environmental considerations relating the reuse and recycle of wastewaters from POTWs containing toxic pollutants from industrial contributions or other sources are important in evaluating the overall impact of toxic pollutants on POTW operations. The primary concern in the reuse of municipal wastewaters is the presence of toxic pollutants which may be deleterious to the environment. The variety of wastewater reuse practices currently employed results in a wide variation of associated environmental considerations. These considerations will also be addressed in Chapter 5.

Pollutants That Inhibit Biological Treatment Processes

A considerable body of information exists in the literature relating to the inhibitory characteristics of many pollutants that may be found in wastewaters treated by a POTW. Classified as materials that are deleterious at trace levels, these pollutants have been shown to decrease biological treatment plant efficiency at low levels, and at higher levels to upset standard treatment processes. Whether a substance is inhibitory depends on a number of factors beyond concentration. Other phenomena, such as synergistic and antagonistic effects as well as biomass acclimation can impact on the degree of inhibition produced by a specific pollutant. Additionally, as a result of the chemical and biological factors which affect the inhibitory impact of a pollutant, a given concentration of that pollutant may be inhibitory to a biological process under a given set of conditions and noninhibitory under another set of conditions. Therefore, it is not surprising that a significant amount of data in the literature are apparently contradictory, in that certain concentrations have been reported as inhibitory, while higher concentrations are reported to be harmless. In most cases, these inversions of effects with concentration are not due to errors of observation, but rather to different conditions surrounding the biological processes, such as those described here.

Synergism

Synergism can be generally characterized as an increase in the inhibitory effect of one substance by the presence of another. Synergism, as well as its opposite, antagonism, is found to be most prevalent when considering combinations of transition metals or heavy metals. The inhibitory effects of these metallic constituents are also enhanced by acidity.

The synergistic effects of metals with acidity is understandable in terms of the chemistry of these metals. The transition and heavy metals tend to be insoluble by hydrolysis in the pH range of sewage influent. They therefore tend to precipitate or adsorb on solids, and interact with polyelectrolytes or various chemical species containing anionic functional groups. Acidity suppresses hydrolysis, and the hydrogen ion competes with the metal ion for adsorption sites on solids or anionic functional groups in solution.

Another type of synergism is encountered with cyanide or other complexing substances which are easily biodegradable. In these cases, it is possible for the microorganism to ingest excessive levels of complexed metal ion and then to destroy by assimilation the complexing substance which is shielding the microorganism from the metal ion. The result is the release of an excessive level of the metal within the organism, upsetting its biological life processes.

Antagonism

Antagonism is the opposite of synergism. It is characterized by a decrease in the inhibitory effect of one substance by the presence of another. The most notable antagonistic effects occur with the combination of metallic and certain anionic pollutants, notably sulfide. Besides sulfide, other ions such as hydroxide (i.e., high pH), chromate, ferrocyanide, phosphate, carbonate, and arsenate will tend to precipitate with metals, thereby reducing the inhibitory effect. Several chelating agents such as EDTA (the disodium salt of ethylenediaminetetraacetic acid) have exhibited antagonistic properties with metals. These chelating agents are used in culturing microorganisms to regulate the level of metals needed to grow bacterial cultures. When chelating agents are present, bacteria thrive in culture solutions containing concentrations of these metals well into the inhibitory range.

Acclimation

In addition to chemical factors, there are significant biological factors which must be considered if the inhibitory effect of pollutants is to be understood. In the activated sludge process, a healthy biomass contains a broad distribution of microorganisms, including many species of bacteria and protozoa. These organisms compete for the available food, oxygen, and nutrients, and grow and reproduce according to the suitability of the aquatic environment to their existence. Whenever the environment changes because of the introduction or omission of a given pollutant, the opportunities for reproduction and growth of different species change, so that the relative sizes of the populations of different species are altered when conditions change.

The history of an activated sludge biomass, therefore, including its distribution of microorganism populations, affects how it will respond when a new pollutant enters its environment. When a new pollutant is introduced, those species which cannot tolerate this substance fail to reproduce and grow, and tend to die off, while more tolerant species consume the food supply and grow and reproduce. When a biomass becomes accustomed to the presence of a normally inhibitory concentration of a substance, it can be characterized as acclimated to that pollutant. Sludge digestion and nitrification do not have the same flexibility of adaptation to changing environmental conditions as do other biological processes. Both nitrification and sludge digestion are biological processes that rely on particular strains of bacteria. As a consequence, when adverse conditions are encountered in these processes, there is no possibility of another organism taking over for the affected strain of bacteria. Consequently, neither nitrification nor sludge digestion are readily acclimated to a new pollutant, and both may be easily upset when new conditions are encountered.

Inorganic Substances

Inorganic species can have varied impacts on POTW operations, ranging from inhibition to beneficial roles as micronutrients. Wastewater entering a POTW may contain any combination of thousands of inorganic compounds present as major or minor constituents. Fortunately, since most inorganic substances dissolved in wastewater are present in ionic form, it is possible to reduce the number of parameters of interest to a smaller list of cations and anions which comprise these dissolved substances. In addition, a few nonionic substances must also be considered.

The list of 129 priority pollutants includes fourteen inorganic pollutants: antimony, arsenic, beryllium, cadmium, chromium, copper, cyanide, lead, mercury, nickel, selenium, silver, thallium, and zinc. These materials were classified as priority pollutants because, among other factors, they were considered toxic. In the literature, ten of the inorganic priority pollutants were also found to be inhibitory to biological treatment processes: arsenic, cadmium, chromium, copper, cyanide, lead, mercury, nickel, silver, and zinc. Table 4–1 summarizes the research data available for the threshold concentrations at which major inorganic substances will inhibit biological processes. For the case of common constituents of water supplies, such as calcium, magnesium, iron, sodium, and sulfate, the threshold concentrations are comparatively high. On the other hand, the threshold concentrations for the priority pollutants are considerably lower, at times smaller by one or two orders of magnitude. In the following sections, detailed information on the inhibitory impact of specific inorganic materials on biological treatment processes is presented. Additionally, general information on the chemical characteristics of these pollutants and their possible sources in a collection system are covered.

Table 4-1 / Threshold concentrations of inorganic pollutants that are inhibitory to biological treatment processes (10).

	Concentration, mg/l[a]			
Pollutant	Activated sludge processes	Anaerobic digestion processes	Nitrification process	References
Ammonia	480	1500		11, 12, 15
Arsenic[b]	0.1	1.6		13, 14
Borate (boron)	0.05–100	2		14, 17, 18, 19
Cadmium[b]	10–100	0.02		13, 15, 20
Calcium	2500			38, 40, 41
Chromium[b] (hexavalent)	1–10	5–50	0.25	4, 14, 15, 21, 26
Chromium[b] (trivalent)	50	50–500		25
Copper[b]	1.0	1.0–10	0.005–0.5	4, 14, 21, 27, 28, 29, 30
Cyanide[b]	0.1–5	4	0.34	4, 13, 14, 22, 35
Iron	1000	5		13, 42
Lead[b]	0.1		0.5	14
Manganese	10			13
Magnesium		1000	50	27, 38, 3, 41
Mercury[b]	0.1–5.0	1365		13, 22, 32, 43
Nickel[b]	1.0–2.5		0.25	4, 22, 47, 49
Silver	5			13, 17, 69,
Sodium		3500		38, 40, 41
Sulfate			500	51, 52
Sulfide		50		46, 52
Zinc[b]	0.08–10	5–20	0.08–0.5	14, 15, 21, 22, 27, 31, 37, 53, 57

[a]Concentrations shown represent influent to the unit processes in dissolved form.
[b]Priority pollutants.

Ammonia. Ammonia (NH_3) is a common chemical substance, which is a gas at room temperature. It is extremely soluble in water and its water solution is used as a household cleaning agent. Ammonia forms a series of salts called ammonium salts. Inasmuch as ammonia and ammonium ion occur together in solution and are transformed from one form to another by shifts in the solution *p*H, the two forms are usually treated as a single substance in most of the wastewater literature. Ammonia occurs naturally in wastewaters as a result of decomposition of nitrogen-containing compounds in sanitary wastes. In addition, ammonia is used in chemical manufacture, water softening, agricultural fertilizers, refrigeration, metal cleaning and other diverse applications.

At low concentration levels, ammonia serves as an important nutrient in biological treatment processes. No adverse effects on oxygen consumption are noted at concentrations of up to 100 mg/l of ammonia, but at excessively high levels (about 480 mg/l) ammonia exhibits inhibitory effects on the activated sludge process (15). At a concentration of 1500 to 3000 mg/l ammonia is inhibitory to anaerobic digestion (11, 12).

Arsenic.   The priority pollutant arsenic is a metallic element which forms cationic salts of the arsenic and arsenous forms and anionic salts of the arsenate and arsenite forms. It is well known for the poisonous properties of its compounds. Arsenic is found at very low concentration levels in most natural waters and is likewise a trace constituent in most foods. Arsenic can enter sewage treatment systems from a number of commercial operations. Arsenic compounds have been used in agricultural pesticides. Other uses associated with its toxic nature are in wood preservatives and in medicines. Arsenic compounds also find commercial use in artists' pigments, glass manufacture, and pyrotechnics. One investigator reports that a level of 0.1 mg/l sodium arsenate (arsenic concentration 0.04 mg/l) showed no effect on oxygen uptake in a biological treatment process, while a level of 1.0 mg/l of this compound depressed oxygen uptake about 50% (13). A 0.1 mg/l concentration of arsenic trichloride (about 0.04 mg/l arsenic) resulted in a significant reduction in sludge digestion efficiency (14).

Borate (and Other Boron Species).   Boron is a light metal. It forms a series of anionic salts, including borate, metaborate, and tetraborate. Boron salts are found widely distributed in natural waters and foods, and therefore are natural constituents of sewage. In addition, boron compounds are constituents of household detergents and medications which also contribute to boron in domestic wastewaters. Boron compounds find application in the manufacture of glass and ceramics, fireproofing, high energy rocket fuels, and in the operation of nuclear reactors. Contributions of boron from these sources are probably small.

Boron concentrations of 0.005 to 0.05 mg/l are reported to have had no effect on microorganism growth (15). However, other investigators report that levels of borate from 0.05 to 100 mg/l interfere with the activated sludge process, e.g., by inhibiting sludge settling and COD removal (14, 15, 17, 18). Boron shock loads of 2 mg/l are reported to adversely affect anaerobic digestion (19).

Cadmium.   Cadmium, which is a priority pollutant, is a transition metal which forms divalent salts. Cadmium salts are somewhat toxic, and are found in domestic water and wastewater sources only at very low levels.

Contributions to POTWs come from industrial or commercial sources, with the principal contributions of cadmium coming from metal plating processes. Cadmium also finds significant use in pigment manufacture, photographic applications, dyeing, and printing.

The results of one research study show that cadmium has no adverse effects on the activated sludge process up to a concentration level of about 1 mg/l (13). However, in the range of 10 to 100 mg/l, cadmium shows various deleterious effects, such as decrease in BOD removal efficiency and reduction in oxygen uptake (13, 15). Another investigator reports that 0.02 mg/l should be considered a threshold concentration for cadmium in digestors (20).

Synergistic effects have been reported between cadmium and zinc, as well as cadmium and manganese. Other heavy metals may also show synergistic effects with cadmium. Sulfide and high pH (8 and above) are strongly antagonistic to cadmium, since they precipitate the insoluble sulfide and hydroxide compounds.

Chromium.   Chromium, which is a priority pollutant, is a transition metal which forms two important series of salts containing the trivalent chromic cation, and hexavalent chromium as part of the chromate anion.

Chromium is contributed to wastewater from numerous industrial and other commercial operations. Especially significant sources are the electroplating and electrofinishing industries. Another significant source is the use of soluble chromates as corrosion inhibitors in cooling towers and insoluble chromates in corrosion resistant coatings. Other sources of commercial discharge of chromium salts include leather tanning, photographic processing, and textile dyeing. Unless stated otherwise, the inhibitory properties outlined below are for hexavalent chromium.

At a concentration level of 0.005 to 0.5 mg/l, chromium has a stimulatory effect on microorganism growth (14). Interference with aerobic biological processes is reported at the 1 mg/l concentration level (14). However, another investigator identifies 10 mg/l as the threshold level for deleterious effects (15, 21). In the concentration range of 1 to 50 mg/l, the published literature is quite confusing and contradictory, reporting results which range from serious interference to insignificant effects (15, 22). In the range of 50 to 500 mg/l, the published results describe various deleterious effects. These include synergistic effects of chromium with acidity, iron, and copper (13, 14, 15, 22). However, slug doses up to 500 mg/l of chromium have also been reported to result in relatively minor disturbances to the activated sludge process (15, 23, 24).

Chromium has also been found to have an inhibitory impact on anaerobic biological processes. Chromium levels of up to 500 mg/l in the plant influent do not affect digestion of the resultant sludge. This result may be due to precipitation of the insoluble chromic hydrous oxide (22). However, the addition of up to 50 mg/l of chromium directly to a sludge digester had a serious effect on digestion (21), and the addition of 500 mg/l to the digester stopped the process completely (22). The effect of trivalent chromium on digestor operation has been reported to be dependent on the digestion period: 500 mg/l trivalent chromium is tolerated in digesters with 20 day digestion periods, 100 mg/l with 17 days of digestion, and 50 mg/l with 14 days (25).

For the nitrification process, it is reported that 0.25 mg/l of chromium is somewhat inhibitory (4), but 2 to 5 mg/l of chromium completely inhibits the nitrification process (5, 15, 26).

Copper.   Copper is also a transition-metal priority pollutant which forms salts and other compounds in two valence forms. Monovalent copper, copper(I), forms what are called cuprous salts, while divalent copper, copper(II), forms what are called cupric compounds. The cuprous compounds are all relatively insoluble and uncommon, and are not significant in water pollution or water treatment chemistry. Most common copper(II) compounds are insoluble, and therefore do not pose a problem to sewage systems in that form. However, cupric nitrate, sulfate, and chloride salts and a number of copper complexes are soluble, and these may pose problems when discharged to sewage treatment plants.

Copper compounds occur naturally in surface and ground waters, usually at relatively low concentration levels, and are natural constituents of domestic drinking water supplies. Copper also enters domestic sewage flows as a result of routine household activities, such as washing and preparing of foods and cleaning of copper utensils. Copper salts discharge to sanitary sewers as a result of corrosion of copper and brass plumbing fixtures and pipes. Corrosion of copper roofing and surface runoff leads to additional discharge of copper to combined sewage systems. Copper compounds are discharged from a number of industrial operations, such as metal cleaning and electroplating operations, engraving, jewelry making, electrical manufacturing, and chemical industrial processes. Finally, algicide and insecticide uses also add copper to industrial wastewater.

Copper is an essential nutritional element for man and lower organisms, and is no doubt essential to the proper operation of biological waste treatment systems at some undetermined trace level. At extremely high concentration levels (about 1000 mg/l and above), practically all forms of living species are destroyed and essentially sterile water results (16).

Synergism between copper and cyanide or heavy metals has been identified. Antagonism has been identified between copper and sulfide and certain chelating agents such as EDTA (27).

For the activated sludge process it has been reported that 1 mg/l should be the threshold limit for continuous feed of copper to the activated sludge process (14, 21, 28, 29, 30). A pilot plant study showed an effect on the process at 1.2 mg/l, while at 10 mg/l a small reduction of plant efficiency of about 4% or less was reported (21, 22, 30, 31).

The combination of 3.6 mg/l of copper with 8.6 mg/l of cyanide caused a serious upset of an activated sludge process (33). Another reference cites 0.1 mg/l as the recommended upper limit of copper ion in sewage feed (28). For slug doses, it has been reported that doses above 50 mg/l for 4 hours show a severe effect (13); a 64 mg/l dose for 4 hours showed a slight effect (22); a 75 mg/l dose for 4 hours "affected" the system (15); and 100 to 400 mg/l showed a severe effect in which plant efficiency dropped to about 50% for 48 hours as a result of the slug dose (22, 28). Another investigator reports that 10 mg/l of copper in the presence of cyanide caused a severe effect (15), while 25 mg/l of copper in the presence of cyanide caused a very severe effect for 24 hours (22). It has also been reported that there is an approximate one-to-one relationship between copper concentration and effluent COD. A 1 mg/l increase in copper will result in a 1 mg/l increase in effluent COD (34).

It should be noted that the severity of effects reported for slug doses of copper does not correspond quite proportionally with the concentrations of these doses. Nevertheless, it may be concluded that slug doses of copper from 50 to 400 mg/l result in serious upset conditions with the activated sludge process.

Data reported on the effect of continuous dosages of from 0.1 to 10 mg/l of copper in influent wastewater on sludge digestor operation vary widely. Two investigators recommend 1.0 mg/l as the maximum concentration of copper in influent wastewater to avoid digestor operating difficulty (14, 35). For direct feed to a combined sludge digestor, a 5 mg/l copper dosage is recommended as an upper limit. For primary sludge digestion, 10 mg/l copper is the recommended upper limit (21). Various copper concentrations in digestors greater than 10 mg/l have been found to be inhibitory (22, 28, 36).

In general, these data show that over the range of copper concentration from 0.1 to 10 mg/l, there are reports of digestor problems attributed to the presence of copper, and other reports of digestor tolerance to these same levels. These discrepancies are no doubt explicable in terms of differences in operating conditions, and antagonistic or synergistic effects.

In the nitrification process, an investigator reports that 0.005 to 0.03 mg/l of copper is stimulatory to nitrifying bacteria, and concentrations above 0.05 mg/l copper were found to be inhibitory (27). Another researcher reports that 0.5 mg/l of copper as copper sulfate is inhibitory to the nitrification process (28).

Cyanide.   The cyanide ion (CN^-) is a pollutant parameter of significant interest in POTW influent and effluent, as well as in rivers, streams, and lakes. The poisonous nature of cyanide is actually associated more with hydrogen cyanide, which generally is more prevalent below pH 7, than with the free cyanide ion. Therefore, cyanide toxicity is directly tied to the pH of the wastewater. Another interesting aspect of its poisonous character is that its toxicity is principally applicable to higher life forms. Microorganisms present in sewage treatment plants can adapt to the presence of cyanide, and metabolize and destroy it even at fairly high concentration levels. An important property of the cyanide ion is that it is a powerful complexing agent and can bind with transition and other heavy metal ions to form metal–cyanide complexes. These complexes exhibit neither the properties of the metal ion nor the cyanide ion, and thus are actually different chemical substances.

Principal sources of cyanide in wastewater are the electroplating, coke, steel, plastics, and chemical industries. The electroplating industry is particularly noteworthy because it combines cyanide wastes with transition and heavy metal ion wastes.

Wastewaters containing 0.01 to 0.05 mg/l of cyanide have been reported to have no deleterious effect on an activated sludge plant (22). However, at levels of 0.3 to 3 mg/l of cyanide, some adverse effects are reported (13). Recommended maximum limits of 0.1 to 2 mg/l of cyanide have been reported in the literature (14, 22), and 5 mg/l of cyanide in raw wastewater has been found to interfere with activated sludge processes (35). A slug dose of 40 mg/l of cyanide upset an activated sludge plant for two days (15). On the other hand, after acclimation, 60 mg/l was tolerated in an activated sludge plant (13) and 200 mg/l was tolerated in a trickling filter plant (15). It has been reported that the toxicity of copper and nickel are enhanced by the presence of cyanide (22, 33), and that cyanide levels of 4 to 100 mg/l upset the sludge digestion process from 4 days to complete retardation (14). Trickling filter operations have been impaired by 30 mg/l of cyanide (37), but 10 mg/l were destroyed by trickling filter operations. Finally, the nitrification process was reported to be inhibited by about 75% by 0.65 mg/l of sodium cyanide (0.34 mg/l of cyanide) (4), and other reports indicate that 2 to 72 mg/l of cyanide as HCN interfered with nitrification, the higher value completely inhibiting the process (14).

In summary, the cyanide ion shows a remarkable range of behavior with biological processes, from interference at low concentrations with non-acclimated systems to tolerance at high levels in acclimated systems.

Iron.   Iron is a transition metal forming two groups of salts, the divalent ferrous salt series and the trivalent ferric salt series. Both the ferrous ion and the ferric ion are precipitated from solution at neutral pH values as hydroxides.

Iron salts are natural constituents of both domestic and commercial wastewaters. While the levels in domestic sewage are generally low, the levels in certain industrial or commercial wastes may be excessive. Major industrial sources include metal pickling and cleaning processes, chemical manufacturing, and electric utilities.

Iron is a necessary element for microbiological growth and its absence causes a reduction in metabolic activity (39, 70). The activated sludge process appears to be rather insensitive to iron concentration except for very high concentration levels. It is reported that 100 mg/l causes little adverse effect, but 1000 mg/l stops oxygen uptake (13).

The sludge digestion process is more sensitive to soluble iron concentration. It is reported that 5 mg/l is a maximum level, as higher iron levels cause interference with the process, due to hydrolysis of the iron and release of acidity (42). The sludge digestion process is very sensitive to pH values outside the optimum range of 6.5 to 7.5, and as a result the inhibitory effect of iron is probably a result of the acidity released (14, 22).

Synergistic effects occur between pairs of transition metal ions present in a wastewater. A specific effect of ion synergism with chromium has been reported and synergism with other metals may also occur (15). Antagonistic effects may be anticipated with sulfide ion and hydroxyl ion (high pH), because of precipitation of the sulfides and hydroxides of iron. An antagonistic effect should be expected with cyanide, since the ferrocyanide complex is very stable and probably noninhibitory to biological processes.

Lead.   Lead, a priority pollutant, is found in natural waters at trace levels as the divalent ion. It occurs in domestic sewage as a result of its presence in the water supply and also as a result of corrosion of lead plumbing. It is present in industrial wastewaters from storage battery manufacture, tetraethyl lead production, and pigment, paint, and cement industries. It also is contributed to wastewater flow as a result of manufacture and use of lead-containing pesticides.

At concentration levels of 0.005 to 0.05 mg/l, lead has no effect on the activated sludge process. A moderate "toxicity" of lead to microorganisms has been reported above 0.1 mg/l and also above 1 mg/l (14). However, in apparently contradictory results, one paper states that a significant effect on oxygen uptake is noted in the presence of lead from 10 to 100 mg/l (15), while a second paper states that no significant change in oxygen consumption is noted at up to 50 mg/l (13). Lead concentrations of up to 0.05 mg/l have been reported to have no effect on the nitrifying bacteria *Nitrosomonas* (27).

Manganese.   Manganese is found in domestic sewage in trace amounts. The significant aqueous form is the divalent ion. Manganese is contributed in the wastewaters from storage battery manufacture, paint manufacture, and chemical manufacture.

It has been reported that a concentration of 7 mg/l of manganese has no adverse effect on the activated sludge process. However, two reports indicate that at 10 mg/l of manganese, a severe adverse reaction occurred with the activated sludge process, and a severe inhibition of oxygen uptake was experienced. Oxygen uptake was completely inhibited at 50 mg/l of manganese (13). Manganese salts in the concentration range of 12.5 to 50 mg/l have the unusual property of stimulating the growth of the microorganism *Nitrosomonas* (15).

Synergistic effects of manganese with zinc and cadmium have been reported (15). It is probable that similar synergistic effects may occur with other transition and heavy metals, and with acidity. Although no data have been found on antagonistic effects with manganese, it may be presumed that hydroxide and sulfide, which can precipitate manganese, will act in this manner.

Mercury.   The priority pollutant mercury forms two series of salts, the monovalent mercurous salts and the divalent mercuric salts. Most mercury salts (of either valence form) are considered to be insoluble or sparingly soluble. However, because of the severe toxicity of mercury to man, fish, wildlife, and lower organisms, even slight solubility poses a substantial threat.

Mercury is found in typical domestic wastewaters at extremely low levels. Commercial contributions occur principally from the chloralkali industry (chlorine–caustic soda manufacture). Other sources include the chemical, drug, herbicide, fungicide, and paper products industries.

At a concentration level of 0.1 mg/l of mercury, one paper reports no effect on the activated sludge process (13); another paper reports a 10% reduction in oxygen uptake (43). A threshold level for adverse effects of mercury on the activated sludge process is reported at about 2.5 mg/l (13). Another investigator reports that at less than 2.5 mg/l, mercury has little effect on aerobic processes, but at above 5.0 mg/l, aerobic processes are definitely inhibited (32). In the concentration range of 1 to 200 mg/l there are numerous reports of different degrees of inhibitory effects on the activated sludge process (13, 15, 44).

A study of mercury behavior in the sludge digestion process indicated that 43 mg/l of mercury in the digester had no adverse effect, while 1365 mg/l had an adverse effect (45).

Nickel.   Nickel is a transition metal and a priority pollutant which forms a series of divalent nickelous salts and trivalent nickelic salts. Only the divalent salts are of interest in wastewater management.

The major source of nickel contribution to wastewaters is from electroplating and related metal finishing processes. Other minor sources of nickel in wastewater arise from corrosion of alloys, dyeing, and printing operations.

There appears to be no significant adverse effect on the activated sludge process from nickel concentrations of less than 1 mg/l (22). However, a threshold effect of nickel on the activated sludge process is reported between 1 and 2.5 mg/l (22). Various adverse effects on the activated sludge process are reported for nickel influent concentrations of 2.5 to 200 mg/l (13, 46, 47). These include reduced oxygen uptake between 10 and 50 mg/l of nickel and interference with solids settling between 2.5 and 10 mg/l of nickel (49).

The data available on sludge digestion in the presence of nickel show little or no effect of nickel (from 10 to 500 mg/l) on the sludge digestion process (13, 15, 22, 31, 36, 46, 48). In view of the sensitivity of microorganisms to nickel ions, it is clear that the reason no effect was noted (except in one instance (42), with up to 500 mg/l of nickel), is because sulfide and perhaps hydroxide precipitate the nickel out of solution. In the absence of adequate sulfide (or sulfate) in the digestor one would expect lower levels of nickel to interfere with the process. Finally, it is reported that a level of 0.5 to 3 mg/l of nickel has an adverse effect, and 10 mg/l has a severe effect on the nitrification process (22, 47).

Silver. The priority pollutant silver is a transition metal which forms a series of monovalent salts. It has few soluble salts. Silver is not normally a significant constituent of domestic wastewaters. Because of its relatively high cost, commercial contributors usually make an effort to recover as much silver as is economically practical. The principal source of silver in wastewaters is the photoprocessing industry. Silver salts are removed from photographic film with sodium thiosulfate (hypo), and the films are rinsed free of the hypo. In large photoprocessing installations, the silver is recovered from the hypo. Traces in rinse water may not be economical to recover, and these residuals result in contributions to the waste load.

Silver ion is extremely toxic to microorganisms, being one of the most reliable disinfectants known. Therefore, it would be anticipated that silver discharges to a POTW could cause interferences with its operation. However, in the case of contributions from the photoprocessing industry, the sterilizing effect of silver is generally not encountered. The reason for this anomaly is that virtually all of the silver released from a photoprocessing plant is in the form of the thiosulfate complex. This complex does not display any of the toxic properties of silver ion. However, free silver ion at the 5 mg/l level causes an 84% inhibition of the activated sludge process (17, 69). At the 25 mg/l level, inhibition is complete (13). In contrast, silver present as the thiosulfate complex, at levels of from 2 mg/l to more than 250 mg/l of silver, had no effect (17,69).

Sulfate. The sulfate radical is a common constituent of natural water supplies and, as a result, is also a common constituent of domestic and commercial wastewater streams. Numerous industries release sulfates; for example, metal cleaning or pickling is carried out with sulfuric acid, which may be neutralized and contributed to a municipal sewer. Similarly, air pollution scrubbers collect abundant quantities of sulfates and sulfites, which may be released to the sewers.

No data have been found on inhibitory effects of sulfate on the activated sludge process but presumably, adverse effects would occur at some elevated concentrations. However, inhibitory effects have been noted in sludge digestion. At greater than 500 mg/l, an adverse effect of sulfate in sludge digestors is reported (11, 51). At a level of 2400 mg/l of sulfate, it is reported that the gas generation was reduced by less than 12% from that without high sulfate level. At a sulfate level greater than 2400 mg/l, there is a report of complete cessation of gas production (46).

Sulfide. The sulfide ion is a common constituent of domestic wastewater, especially when anaerobic conditions occur. Sulfides may be discharged in wastes from petroleum refining, the leather tanning industry, and chemical manufacturing industries.

Excessive levels of sulfide interfere with the activated sludge process by depleting the dissolved oxygen transferred in the aeration process. One investigator reports that 25 to 50 mg/l of sulfide is tolerable for about one week (15). Sulfide is beneficial to the anaerobic digestion process at low concentrations because of its ability to precipitate transition and heavy metals out of solution. At higher concentrations, and subject to conditions in the digestor, sulfide becomes inhibitory. Reports of tolerance to sulfide in the sludge digestion process include the following: 50 to 100 mg/l of sulfide can be tolerated (38); 200 mg/l causes less than 12% loss in gas generation (46); and up to 200 mg/l of sulfide can be tolerated (38). Sulfide concentrations of 150 to 200 mg/l in the digestor feed can reduce gas production considerably (46, 52).

Reports of inhibitory effects include the following conclusions: more than 50 mg/l of sulfide reduces gas production considerably; 100 mg/l of sulfide causes from a 33% to 50% loss in gas generation; no methane forms with more than 165 mg/l of sulfide; 200 mg/l of sulfide causes an 80% loss of gas generation; more than 200 mg/l of sulfide is quite toxic, causing complete cessation of gas generation; and 400 mg/l of sulfide causes 95% loss of gas generation (45). It is clear that the effect of a given level of sulfide ion may be quite variable, depending on specific process conditions. The reported contradictory effects of sulfide an anaerobic digestion points to the dependence of inhibition or tolerance on other factors.

Zinc. Zinc, a priority pollutant, is a transition metal which forms a series of divalent salts. It has amphoteric behavior, i.e., it forms zinc cations and zincate anions. Zinc has a rather widespread occurrence, and is a normal constituent of domestic wastewaters. Zinc is contributed to sewage flow from a number of industries, including electroplating, dye and pigment manufacture, rubber processing, and electrical generation.

A number of reports cite tolerance of the activated sludge process to zinc up to concentration levels of 10 mg/1 (13, 31, 53). There is a report of an adverse effect in the 0.08 to 0.5 mg/1 range and a report that the threshold level is between 5 and 10 mg/1 (15, 54). Additional reports of inhibitory effects are given over the range of 20 to 160 mg/1 (14, 15, 22, 53, 55). Synergistic effects are noted between 1 mg/1 of zinc and 10 mg/1 of cadmium, between 10 mg/1 of zinc and 100 mg/1 of manganese, and between 10 mg/1 of zinc and 10 mg/1 of cadmium.

Two references cite no adverse effects of zinc with the sludge digestion process for 10 mg/1 and 10 to 20 mg/1 of zinc, respectively (48, 53). However, two references cite adverse effects of zinc on the sludge digestion process at 20 mg/1 and 1000 mg/1 respectively (14, 22). Other investigators indicate 5 mg/1 zinc as the upper limit to prevent decreases in digestor gas production (37, 42), and 10 mg/1 as the highest continuous dosage that will allow satisfactory digestion (21).

Adverse effects on solids settling have been reported for 7.5 and 15 mg/1 slug doses of zinc solution over a half-hour period (15), and inhibition of nitrifying bacteria by 0.08 to 0.5 mg/1 of zinc has also been reported (27).

Organic Substances

Organic chemicals may be characterized as those compounds made up of carbon in combination with hydrogen, oxygen, nitrogen, sulfur, or phosphorus. The bulk of the priority pollutants (114 in total) are organic chemicals, and many of these materials can be found in sewage. Although many organic chemicals are biodegradable and therefore compatible with standard biological treatment processes, some organics have been known to upset POTWs. Anaerobic digestion and nitrification processes appear to be most susceptible to upset by organic pollutants, particularly chlorinated hydrocarbons (29). Autotrophic bacteria, which are involved in nitrification, are more readily inhibited than are the heterotrophic bacteria, which are involved in the oxidation of carbon compounds. Organic sulfur compounds, especially those with sulfur–carbon–nitrogen linkage, are inhibitors of nitrification (29). The significant organic constituents which have been identified as having inhibitory effects are discussed below, and are summarized in Table 4-2.

In the following sections, the inhibitory impact of specific organic pollutants on POTW operations will be discussed.

Alcohols.   An alcohol is a hydrocarbon in which one of the hydrogens is replaced by a hydroxyl (—OH) group. The lower-molecular-weight alcohols are relatively polar substances and are completely miscible with water. With increasing size of the hydrocarbon group the alcohols become increasingly insoluble in water. Polyhydroxy alcohols contain more than one hydroxy group per molecule. One of these, ethylene glycol, is used extensively as an antifreeze. Alcohols may be found in the effluents from the pharmaceuticals, alcoholic beverages, antifreeze chemicals, and plastics manufacturing industries. One investigator reports that 19.5 mg/l of allyl alcohol inhibits nitrification by 75% (4). Table 4-3 summarizes the impact of some alcohols on araerobic digestion.

Phenols.   Phenols are aromatic hydrocarbons with a hydroxyl (—OH) group substituted for a hydrogen in the ring. Cresols are phenols with a methyl group substituted for a second hydrogen in the benzene ring. Domestic wastewaters do not generally contain significant amounts of phenols. Phenols can cause significant problems if present in domestic water supplies, since the chlorophenols that may be produced during chlorination can be odorous.

Phenols are basic organic chemicals used in many industrial syntheses, including plastics production, dyes, and pharmaceuticals. Phenolic wastewater is also produced as a byproduct of petroleum refining. The specific compound phenol C_6H_5OH and miscellaneous other phenolic compounds, are priority pollutants.

There is an extremely diverse reaction to phenolic wastes in the activated sludge process, depending upon whether the sludge has been acclimated. Relatively small amounts of phenol can be inhibitory to unacclimated sludge. However, with acclimation, and use of the complete mixing mode of operation, high concentrations of phenol can be tolerated.

Phenol slug doses of 200 mg/l can deactivate aerobic treatment plants by killing the biomass (7). One investigator reports a progressive inhibition of nitrification between 4 and 10 mg/l of phenol or cresol. 5.6 mg/l of phenol has been found to inhibit nitrification by 78% (4). Cresol at concentrations of 4–5 mg/l inhibits nitrification (56). 12.6 mg/l of *o*-cresol, 11.4 mg/l of *m*-cresol, and 16.5 mg/l of *p*-cresol were found to inhibit nitrification by 75% (4). A particular bacterium found in the activated sludge biomass (*bacillus cereus*) is capable of metabolizing phenol. Concentrations of up to 1000 mg/l of phenol are not harmful to this species (64).

Chlorinated Hydrocarbons.   Hydrocarbons in which one or more hydrogen atom is replaced by a chlorine atom can be classified as chlorinated hydro-

Table 4-2 / Threshold concentrations of organic pollutants that are inhibitory to biological treatment processes.

	Concentration, mg/l[a]			
Pollutant	Activated sludge processes	Anaerobic digestion processes	Nitrification processes	References
Alcohols				
allyl		100	19.5	4, 51
crotonyl		500		51
heptyl		500		51
hexyl		1000		51
octyl		200		51
propargyl		500		51
Phenols				
phenol	200		4–10	7, 56
creosol			4–16	4, 56
2-4-dinitrophenol			150	4
Chlorinated hydrocarbons				
chloroform		10–16		51, 57
carbon tetrachloride		10–20		59, 58, 29, 60
methylene chloride		100–500		60, 59
1-2-dichloroethane		1		29
dichlorophen[b]		1		29
hexachlorocyclohexane		48		29
pentachlorophenol[b]		0.4		29
tetrachloroethylene		20		29
1, 1, 1-trichloroethane		1		29
trichloroethylene		20		29
trichlorofluoromethane[b]		0.7		29
trichlorotriflouroethane (Freon)		5		29
allyl chloride			180	4
dichlorophen			50	4
Organic nitrogen compounds				
acrylonitrile		5		29
thiourea			0.075	4, 29
thioacetamid			0.14	4, 29
aniline			0.65	4
trinitrotoluene (TNT)	20–25			54
EDTA	25		300	29, 61
pyridine			100	56

Table 4–2 / *(Continued)*

Pollutant	Concentration, mg/l[a]			
	Activated sludge processes	Anaerobic digestion processes	Nitrification processes	References
Surfactants				
Nacconol	200			62
Ceepryn	100			62
Miscellaneous organic compounds				
benzidine	500	5		63
thiosemicarbazide			0.18	4
methyl isothiocyanate			0.8	4
allyl isothiocyanate			1.9	4
dithiooxamide			1.1	4
potassium thiocyanate			300	4
sodium methyl dithiocarbamate			0.9	4
sodium dimethyl dithiocarbamate			13.6	4
dimethyl ammonium dimethyl dithiocarbamate			19.3	4
sodium cyclopentamethylene dithiocarbamate			23	4
piperidinium cyclopentamethylene dithiocarbamate			57	4
methyl thiuronium sulfate			6.5	4
benzyl thiuronium chloride			49	4
tetramethyl thiuram monosulphide			50	4
tetramethyl thiuram disulfide			30	4
diallyl ether			100	4
dimethylparanitrosoaniline			7.7	4
guanidine carbonate			19	4
skatole			16.5	
			7.0	4
strychnine hydrochloride			175	4
2-chloro-6-trichloromethylpyridine			100	4
ethyl urethane			250	4
hydrazine			58	4
methylene blue			100	4
carbon disulfide			35	4
acetone			840	4
8-hydroxyquinoline			73	4
streptomycin			400	4

[a]Concentrations shown represent influent to the unit process, except as noted.
[b]Concentration represents total plant.

Table 4–3 / Effect of some alcohols on anaerobic digestion (51).

Alcohol	Use	Water solubility	Inhibitory concentration, mg/l
Allyl (CH$_2$=CH—CH$_2$OH)	plastics	miscible	100
Propargyl (CH=C—CH$_2$OH)		miscible	50
Crotonyl (CH$_3$—CH=CH—CH$_2$OH)		partly soluble	500
Hexyl (C$_6$H$_{13}$OH)	antiseptic	slight	1000
Heptyl (C$_7$H$_{15}$OH)		slight	500
Octyl (C$_8$H$_{17}$OH)	perfume	insoluble	200

carbons. These materials do not occur in nature but can be found in domestic water supplies as a result of chlorine disinfection and the consequent chlorination of trace hydrocarbons. Chlorinated hydrocarbons are known for their persistance in the environment and are considered especially dangerous because of their tendency to accumulate in the tissues of higher life forms. As a result, many chlorinated hydrocarbons, particularly pesticides, have been banned from environmental use, or their use has been curtailed. Numerous chlorinated hydrocarbons have also been classified as priority pollutants.

Chlorinated hydrocarbons are employed in a broad range of industrial applications, including use as solvents and degreasing agents, agricultural chemicals, disinfectants, dry cleaning agents, soaps and shampoos, wood preservatives, propellants, and refrigerants, as well as in chemical manufacturing. The inhibitory properties of several important chlorinated hydrocarbons are detailed below.

Chloroform. The priority pollutant chloroform (CHCl$_3$) is a low-boiling-point liquid which is only slightly soluble in water. It is used as a solvent for fats, oils, rubber, alkaloids, waxes, resins, as a cleansing agent, in fire extinguishers, and in the rubber manufacturing industry.

Continuous doses of chloroform at 16 mg/l or more in the raw sludge feed has caused inhibition of anaerobic digestion. Continuous doses at concentrations of 10 and 11 mg/l produces a noticeable drop in gas yield (51, 57).

In another study, it was found that at 1.5 mg/l of chloroform there is no inhibition of sludge digestion, while at 14.9 mg/l inhibition is complete (59). One investigator reported slight reduction in digestor gas production caused

**Table 4-4 / Impact of chloroform shock
loads on digestor gas production (57).**

Shock dose of $CHCl_3$ in raw sludge feed, mg/l	Average inhibition, %
1	3.1
5	10
10	16.9
16	42.3
20	54.3

by chloroform at a concentration of 0.1 mg/l. Also included in this report is a reference to an investigation that reported a 50% gas reduction due to 0.96 mg/l of chloroform (60). The effect of shock dosing with chloroform as determined in laboratory digestors is summarized in Table 4-4. The table shows the average percentage inhibition of gas production for each shock dose in the raw sludge.

Carbon Tetrachloride.   Carbon tetrachloride is also a priority pollutant which is a colorless, nonflammable liquid, insoluble in water, with a characteristic odor. It is very toxic to humans. It is a general purpose solvent with broad industrial applications. Carbon tetrachloride has been used as a fire extinguisher, a drying agent, a chemical intermediate, and as an exterminating agent.

Research shows that carbon tetrachloride is primarily inhibitory to anaerobic sludge digestion processes. At levels of 10 mg/l in sludge, carbon tetrachloride has been found to be toxic to anaerobic digestion (29). Another investigator reports that 2.2 mg/l reduced methane production by 50% (60), and further research showed that 100% inhibition of gas production during anaerobic digestion required a carbon tetrachloride dosage of 16 mg/l (58). Table 4-5 summarizes the inhibition of sludge digestion processes that might be expected from various levels of carbon tetrachloride.

Methylene Chloride.   The priority pollutant methylene chloride (CH_2Cl_2) is a colorless, nonflammable liquid which is slightly water soluble. It is used as a solvent for cellulose acetate, as a degreasing agent, a cleaning fluid, and an anesthetic.

**Table 4-5 / Impact of carbon
tetrachloride on sludge digestion (59).**

Concentration level, mg/l	Inhibition, %
0.8	0
7.9	40
19.7	90
159.4	100

Table 4-6 / Impact of methylene chloride on sludge digestion (59).

Concentration, mg/l	Inhibition, %
1.3	0
3.3	40
9.9	60
132.6	80
530.4	100

A methylene chloride concentration of 1 mg/l in the influent to a sewage plant is reported to inhibit sludge digestion (29). Methane production in anaerobic digestion was also reduced by 50% in the presence of 100 mg/l of methylene chloride (60). In Table 4-6 the impact on anaerobic digestion of methylene chloride at various dosages in the raw sludge is summarized.

Chlorobenzenes.   Chlorobenzenes have varied commercial uses including applications as solvents and chemical intermediates. In Table 4-7, the results of a series of batch experiments to determine the impact of chlorobenzenes on digestor gas production are summarized. Chlorobenzene and dichlorobenzene are both toxic materials classified as priority pollutants.

Miscellaneous Chlorinated Hydrocarbons.   Table 4-8 presents the concentrations of various chlorinated hydrocarbons that have exhibited inhibitory effects on anaerobic treatment processes. Another investigator reports a

Table 4-7 / Impact of chlorobenzenes on sludge digestor gas production (60).

Material	Time from start of test, days	Concentration in digestor contents, % wt/wt dry sludge solids, for given reduction in gas production			
		20% reduction	35% reduction	50% reduction	80% reduction
Chlorobenzene	2	1.3	1.7	2.1	3.5
	4	1.0	1.4	1.9	3.4
	6	0.94	1.3	1.7	3.0
Orthodichlorobenzene	2	1.1	1.6	2.4	5.4
	4	0.88	1.4	2.0	4.7
	6	0.73	1.2	1.6	3.8
Paradichlorobenzene	2	2.1	2.6	3.0	8.0
	4	1.8	2.4	3.1	7–8.0
	6	1.4	2.1	2.7	5.3
Certi-chlor[a]	2	0.7	1.6	2.5	4.9
	4	0.62	1.3	2.1	4.9
	6	0.54	1.1	2.0	4.3

[a]A proprietary material similar to ortho-dichlorobenzene and consisting largely of that material.

Table 4-8 / Toxicity of some chlorinated hydrocarbons to anaerobic digestion (29).

Chlorinated hydrocarbon	Inhibitory concentrations, mg/l	
	In sewage	In sludge
1, 2-Dichloroethane		1
Dichlorophen	1	
Hexachlorocyclohexane		48
Pentachlorophenol	0.4	
Tetrachloroethylene		20
1, 1, 1-Trichloroethane		1
Trichloroethylene		20
Trichlorofluoromethane	0.7	
Trichlorotrifluoroethane (Freon)		5

slight reduction in digestor gas production as a result of a concentration of 0.1 mg/l of 1,1,1-trichloroethane (60). In another experiment, a 75% reduction in nitrification was reported in the presence of 180 mg/l of allyl chloride (4).

Agricultural Chemicals.   Insecticides, pesticides, and herbicides are manufactured from chlorinated hydrocarbons and from organophosphorus compounds. Some of these materials are damaging to sewage treatment processes and to fish and wildlife in very small concentrations (DDT, aldrin, dieldrin). Whereas most other chlorinated hydrocarbons enter the wastewater stream as a result of industrial manufacturing processes, agricultural chemicals usually enter wastewater and natural streams by means of runoff.

One laboratory test revealed that two pesticides, aldrin and simazine, were not inhibitory to the growth of the nitrifying bacterium *Nitrobacter,* whereas five other pesticides, including chlordane, heptachlor, lindane, CIPC, and DDD prevented growth. Heptachlor was the most deleterious compound (65).

Organic Nitrogen Compounds.   Many organic compounds contain nitrogen, commonly in the form of a carbon–nitrogen bond. Because of the numerous nitrogen oxidation levels, the chemistry of organic nitrogen substances is quite diverse. At its lowest oxidation state nitrogen forms amines. The most common amine is ammonia. Amines are classified according to the number of hydrocarbon groups attached to the nitrogen atom. Thus, one substitution gives a primary amine, two substitutions give a secondary amine, three yields a tertiary amine, and four produces quaternary ammonium salts. Amines are generally soluble in water, and ionize, exhibiting varying pH-dependent properties. At its highest oxidation state, nitrogen forms nitro-compounds. These materials are chemically unstable and are easily, and sometimes violently reduced (trinitrotoluene, TNT, is a typical example). Nitrogen exists at intermediate oxidation states, each producing compounds with differing properties.

While some nitrogen compounds are compatible with biological treatment, others are inhibitory. Some nitrogen is required for satisfactory treatment plant operation, and a nitrogen source must be added to those trade wastes which are nitrogen deficient. Some organic nitrogen compounds are completely biodegradable, whereas others are toxic to wildlife and some are carcinogenic to man. Organic nitrogen is contributed to sewage from domestic wastewaters which contain complex proteinaceous materials, from agricultural runoff, and from a variety of industrial operations. Specifically, this class of compounds is used in the textile, dye, pharmaceutical, plastics, varnish, perfume, insecticide, tanning, synthetic fiber, and solvents industries.

One investigator reports that 5 mg/l of acrylonitrile in sludge has an inhibitory effect on anaerobic digestion (29), while another investigator reports that more than 20 mg/l of acrylonitrile in sludge is not harmful to anaerobic digestion (68). In other research, 0.075 and 0.076 mg/l of thiourea have been reported by two investigators as inhibiting nitrification (4, 29), and 0.14 and 0.18 mg/l of thioacetamide have been reported as inhibiting nitrification (4, 29). Aniline at a concentration of 0.65 mg/l was also found to inhibit nitrification by 75% (4). At concentrations of 20–25 mg/l of TNT, aerobic processes were severely inhibited (54).

Increasing amounts of EDTA adversely effect settleability of secondary sludge and increases surface scum. At more than 10 mg/l there is a decreased utilization of oxygen. At more than 25 mg/l there is a toxic effect on coliform bacteria (61). A concentration of 300 mg/l of EDTA has been found to inhibit nitrification by 75% (4).

Pyridine and the methylpyridines which occur in coke oven liquors are varied in their ability to inhibit nitrification. The impact of these materials is summarized in Table 4–9.

Surfactants. This group of chemicals comprises those substances which have surface-active properties. This includes synthetic detergents, emulsifiers, foaming agents, and wetting agents. These substances comprise normal constituents in domestic sewage, mostly associated with laundry detergents and commercial cleaning formulations. In addition, they are contributed by numerous industrial and commercial sources, including commercial laundries, wool scouring plants, dyeing, and rubber processing.

Surfactants may interfere with operation of a POTW in a number of ways. The surfactant may be inhibitory to biological processes because of its chemical properties. The surfactant may also interfere through various

Table 4-9 / Impact of pyridenes on nitrification (56).

	Concentration, mg/l	Inhibition of nitrification, %
Pyridine	100	100
4-Methylpyridine	100	100
3-Methylpyridine	100	no effect
2-Methylpyridine	100	40

physical effects by causing excessive foaming, interfering with oxygen transfer or dispersing the biomass floc and causing loss of solids.

Laboratory tests with the anionic surfactant Nacconol (at 100 mg/l) showed a stimulatory effect on the activated sludge process. At concentration levels greater than 200 mg/l, inhibitory effects were noted. These effects were worse at low pH (about 5) and low sludge loadings (62). In another laboratory test with the cationic surfactant Ceepryn (cetylpyridinium chloride), 100 mg/l of this material suppressed oxygen uptake. The effect was more deleterious at high pH (about 9) (62). In a pilot plant study, it was noted that 10 mg/l of an alkylbenzene sulfonate (ABS) surfactant had a negligible effect on the activated sludge process. It was reported that higher concentrations can be very inhibitory (66). In another test, 10 and 20 mg/l levels of ABS caused minor losses (5–8%) in BOD removal efficiency in the activated sludge process. In the same study, a linear alkyl sulfate (LAS) at 10 and 20 mg/l caused a loss of only 1–2% BOD removal in the activated sludge process (67). A laboratory study determined that a surface-active agent at concentration levels from near zero to 50 mg/l interfered with oxygen transfer to water (50). Digestion is likely to be seriously affected if the detergent concentration exceeds 2% of the weight of suspended solids. Initiation of digestion processes may be difficult at a detergent concentration of only 1% (29).

Miscellaneous Organic Chemicals

Benzidine. A laboratory experiment is reported which showed that 500 mg/l of benzidine inhibited oxygen uptake for 144 hours. At 5 mg/l sludge activity decreased (63).

t-Butyl Borane. In laboratory tests 500 mg/l of *t*-butyl borane was found to be inhibitory to activated sludge processes.

Formaldehyde. A level of 500 mg/l of formaldehyde was found to be stimulatory to activated sludge processes.

Benzene, Toluene, Xylene.   Benzene and xylene at the 1000 mg/l level are reported to seriously retard sludge digestion. Toluene at the 500 mg/l level has no appreciable effect on the process.

REMOVAL AND PASSTHROUGH OF TOXIC POLLUTANTS

Introduction

In addition to inhibiting normal POTW operations, toxic pollutants contributed to the sewer by industries may pass through the system adversely, impacting on receiving stream quality. Conversely, within the POTW, toxic pollutants that were not considered in the design of the treatment plant may be removed via incidental mechanisms. Although removal might appear to be beneficial, it often occurs at the expense of sludge quality by concentrating in plant residues. Therefore, in either case, the presence of toxic pollutants in a POTW influent creates a potential for adverse environmental impact, even if treatment plant processes are not directly impaired.

Mechanisms for Toxic Pollutant Removal

In general, toxic pollutant removal mechanisms within a POTW can be classified into three broad categories: physical, chemical, and biological. The exact combination of these phenomena affecting any particular priority pollutant depends largely on the nature of the pollutant itself. For the purpose of addressing toxic pollutant removal in POTWs, toxic pollutants may be grouped into several general classifications:

1. Metals such as cadmium, chromium, copper, nickel, lead, etc.;
2. Volatile organics that are also relatively biodegradable, such as straight chain aliphatics;
3. Volatile organics that are refractory, such as aromatics;
4. Semivolatile organics.

For each class of toxic pollutant, a different set of removal mechanisms are applicable.

Physical Removal Mechanisms

There are three types of physical removal of toxic pollutants in POTWs: removal as a solid with other suspended solids, adsorption onto suspended solids with subsequent removal, and atmospheric stripping.

Removal of toxic pollutants with suspended solids in primary sludge is most prevalent for the toxic metals. Combination of the heavy metals with alkalinity or sulfide will produce insoluble species that settle out of raw sewage simultaneously with other wastewater solids.

It is often the case that the aqueous environment in sewers is chemically reducing, creating sulfides which combine with toxic metals. As a consequence, metal sulfides may be formed which are sparingly soluble and reach the treatment plant as solids which are removed in grit chambers or primary clarifiers. Similarly, the sulfide present in anaerobic digestors often precipitates metals creating additional removal.

Adsorption onto solid surfaces provides an additional removal mechanism for some organic and complexed inorganic or metallic toxic pollutants. If an organic material is insoluble in water, slightly soluble, or hydrophobic, the organic pollutant may preferentially adsorb on solid surfaces. In raw sewage, both suspended solids and floatables (greases) may be available for adsorption. Therefore, when scum or primary sludge is removed, organic priority pollutants, which may be concentrated in these materials, may also be removed. Likewise, metals may complex with organic materials in sewage suspended solids, being removed along with those solids in sludge streams.

A significant proportion of the organic priority pollutants are relatively volatile. It has been postulated that during aeration some of these materials may be air stripped and subsequently released to the atmosphere. This phenomenon might account for some observed removals of biologically refractory volatile organics, especially aromatic species, in activated sludge plants. However, the degree to which volatile organics are stripped during activated sludge treatment is dependent on divergent factors such as vapor pressure of the volatile species, temperature, and competing processes such as other removal phenomena.

Chemical Removal Mechanisms

Chemical removal of toxic pollutants generally is applicable to organic materials which may come in contact with strong oxidants in a POTW. Most commonly, chlorine used for disinfection or odor control reacts with organic pollutants. At times, the organic species are simply chlorinated, sometimes creating toxic materials. However, removal may occur when the oxidation of the organic material goes to completion, destroying the toxic pollutant and forming carbon dioxide and water. Chlorine is not the only oxidant used within a POTW which could destroy organic pollutants. Hydrogen peroxide, which is sometimes used to control filamentous bulking, or ozone, which is gaining acceptance as a disinfectant, can also provide beneficial oxidation and removal of organic pollutants. Oxygen from aeration processes or in pure oxygen activated sludge systems may also contribute to the oxidation of some materials.

Biological Removal Mechanisms

Under the proper conditions, organic toxic pollutants may be biologically removed from sewage by acting as substrate for organisms in a treatment plant's biomass. Aliphatic compounds are generally more amenable to breakdown in biological systems than are aromatic compounds, straight chain aliphatics being most easily degraded. In order for an organic pollutant to compete as a food source with normal organic constituents in sewage (carbohydrates, proteins, fatty acids, etc.) certain conditions should be maintained, such as acclimation to the possible toxic effects of the pollutant. If these conditions can be met, significant removal by biodegradation can occur.

A sometimes overlooked mechanism for the removal of inorganic priority pollutants via biological processes is the uptake of trace quantities of normally toxic inorganic pollutants as micronutrients. These materials may find their way into the biomass as a result of being complexed and encapsulated in a material that is consumed by cells.

Influent Concentrations of Toxic Pollutants

Metallic toxic pollutants can be present in a POTW's influent as a result of industrial contributions or from uncontrollable diffuse resources. In either case, excluding occasional accidental spills or industrial pretreatment plant upsets, the levels of these metallic pollutants observed in a POTW influent are relatively low. In a recent survey of 103 POTWs (71), located in various regions of the United States, influent metals data were collected. The study showed that although metals concentrations can vary by as much as three orders of magnitude in a POTW's influent, the concentration of the pollutant metals, with the exception of zinc, is rarely above 1 mg/l. Further, while the majority of the 103 plants covered in the study accepted significant industrial flow, the few purely domestic plants studied showed metals concentrations close to the overall data base median. This indicates that a significant source of metallic priority or toxic pollutants entering a POTW are nonindustrial and presumably not controllable, and that observed fluctuations in influent metals levels are most probably a result of industrial contributions. However, under any circumstances, the observed influent metals levels are low, which is probably a result of dilution in the collection system.

In comparison to metallic toxic pollutants, data on influent concentrations of toxic organic pollutants is slim. In a recent study (72), two POTWs were sampled over one-week periods. The collected samples were analyzed for the 129 priority pollutants as well as selected conventional and nonconventional pollutants. The two POTWs studied were both activated sludge plants, but the quantities of industrial wastewaters contributed to each plant were significantly different. Plant A received approximately 30% of its hydraulic loading from industrial sources. This flow accounted for about 70% of the plant's organic loading. In contrast to this, plant B almost exclusively treated domestic and commercial wastewater, with industrial flow accounting for only a small percentage of the total plant flow.

The analytical results for this two-plant study showed that there is a significant difference in the incidence and concentration of organic priority pollutants in plant influent between POTWs with high industrial flow and those with purely domestic plants. At plant A the organic pollutants with the highest concentration in the influent were benzene, 1,1,1-trichloroethane, chloroform, ethylbenzene, *bis*(2-ethylhexyl)phthalate, tetrachloroethylene, toluene, and trichloroethylene. All of these organic pollutants consistently were measured at levels in excess of 25 μg/l. On the other hand, at plant B no organic priority pollutant occurred at an average of over 20 μg/l in the plant's influent. The organic pollutants which were present in the highest concentrations, however, were benzene, methylene chloride, and *bis*(2-ethylhexyl)phthalate. Overall, fewer priority pollutants were detected in the plant B influent as compared with plant A. In total, 52 toxic organic pollutants were found in the plant A influent, while in the plant B raw wastewater only 33 were detected. Similarly, 18 organic priority pollutants were measured at above the detection limit at plant A, but at plant B only five organic priority pollutants were found at above detectable levels. It was also found that the more industrial plant A influent contained 21 organic priority pollutants absent in the plant B influent, but only two organic priority pollutants were found exclusively in the plant B raw wastewater. Concentrations of organic priority pollutants were also higher in the plant A influent. A total of 20 organics found in both raw wastewaters had higher average concentrations in the plant A influent. Conversely, only six toxic organic pollutants common to both raw wastewater streams were more concentrated in the plant B influent. It is also interesting to note that 13 of the 20 organics found at higher average concentrations in the predominantly industrial plant A influent are solvents.

Nine metallic priority pollutants were detected in the influents to both plants. Seven of the nine metallic pollutants were found to have higher average concentrations in the plant A influent. Mercury had the same average concentration in both influents, and only zinc was measured at a higher level in plant B as compared with plant A influent. It should be noted that the traditional pollutant parameters (BOD, COD, TOC, residue, etc.) were also consistently higher in the plant A raw wastewater as compared to the plant B influent.

Toxic Pollutant Removal Efficiencies

In a survey of 269 POTWs (10) which compiled self-monitoring data generated by plant personnel, removal efficiencies for selected toxic metals and
phenol were presented for primary, trickling filter, and activated sludge
plants. Table 4-10 summarizes these data and shows that the removal of
metals in primary plants was generally low, with cadmium and nickel having
the lowest removals. Mean effluent concentrations were 14 $\mu g/l$ for cadmium and 165 $\mu g/l$ for nickel. Removal of chromium, lead, copper, and
mercury was somewhat higher, while zinc removals were the highest for
toxic metals in primary plants. Mean effluent concentration is 550 $\mu g/l$ for
zinc.

In biological treatment plants (activated sludge and trickling filter) cadmium and nickel were also removed least, with chromium, lead, copper, and
mercury removals being slightly higher. Mean effluent concentrations of up
to 30 $\mu g/l$ for cadmium and 182 $\mu g/l$ for nickel were reported. Zinc and
copper had a relatively high percentage of removal in biological treatment
plants. Mean effluent concentrations of up to 277 $\mu g/l$ for zinc and 113 $\mu g/l$
for copper were reported. These data would tend to indicate that zinc and
copper are most susceptible to removal in conventional treatment facilities,
while cadmium, nickel, and manganese are the least susceptible of the metals
to removal.

In the two-plant study (72), during the week of sampling, plant A, which
had a substantial industrial flow, achieved good removal for priority pollutant metals that were initially present in detectable amounts. Antimony,
arsenic, beryllium, selenium, and thallium were never detected above their
detection limits in influent or effluent samples, and percentage removals
could not be calculated. Chromium and copper both were reduced to less
than 50 $\mu g/l$ (90% and 86% removal, respectively). Cadmium, nickel, and

Table 4-10 / Toxic pollutant removal data: survey of 269 POTWs (10).

Parameter	Primary plants		Trickling filter plants		Activated sludge plants	
	Mean removal, %	No. of plants	Mean removal, %	No. of plants	Mean removal, %	No. of plants
Cadmium	8	31	20	35	17	44
Chromium	26	36	37	48	46	51
Lead	24	34	37	41	39	49
Mercury	27	21	30	20	39	34
Copper	26	44	54	49	57	63
Nickel	6	28	21	32	20	44
Zinc	31	38	46	52	58	58
Phenol	38	2	50	12	69	16

zinc were removed somewhat less effectively (59–65% each, on average). Lead and silver were both reduced to below their detection limits, precluding accurate calculation of percentage removals. Nine organic priority pollutants were detected in plant A's influent at an average of over 10 μg/l. Eight of the nine (benzene, 1,1,1-trichloroethane, chloroform, ethylbenzene, *bis*(2-ethylhexyl)phthalate, tetrachloroethylene, toluene, and trichloroethylene) were reduced by a minimum of 50%. Only phenol was poorly removed. Carbon tetrachloride and 1,1,2-trichloroethane were each measured at an average concentration of several micrograms per liter (μg/l) in the influent, but were not measured in any effluent sample.

Plant B, with a relatively low industrial flow, had relatively low influent metal concentration with correspondingly low effluent levels. Antimony, arsenic, beryllium, selenium, and thallium were not measured above their detection limit in either influent or effluent samples. Cadmium and silver were both reduced from several μg/l to below their detection limits. Cadmium, copper, and zinc were reduced effectively, between 60% and 81%. Lead and nickel were removed less effectively. Organic priority pollutants at plant B occurred at such low average concentrations that percentage removal data were not meaningful.

Sludge Concentrations

For plant A, most of the metals occurred in high concentrations in both the primary and secondary sludge. Cadmium, copper, lead, nickel, and zinc were each found in primary sludge at concentrations over 100 times greater than their concentration in the influent. Antimony, arsenic, and beryllium, which were never measured above their detection limits in the influent, were all measured in the primary sludge. Chromium and cyanide were found in the primary sludge at 30 to 50 times their influent concentration. Chromium had a higher than expected concentration in the secondary sludge.

Several organic priority pollutants that were detected at very low concentrations in the influent accumulated in the primary or secondary sludge. Among them were acenaphthene (0–1 μg/l average in the influent vs. 169 μg/l in the primary sludge), dichlorobromomethane (less than 0.5 μg/l vs. 57 μg/l), 1,2-benzanthracene (less than 1 μg/l vs. 479 μg/l), 3,4-benzofluoranthene (not detected vs. 675 μg/l), fluorene (less than 3 μg/l vs. 313 μg/l), and pyrene (less than 3 μg/l vs. 757 μg/l).

Data from plant B indicated the same general trends. Chromium, copper, lead, nickel, and zinc were found in the combined sludge at approximately 100 times their concentration in the influent. Arsenic, cadmium, cyanide, mercury, and silver also accumulated in the sludge, but occurred at overall lower levels. Antimony, beryllium, selenium, and thallium, which were never measured above their detection limit in the influent, were all found in the sludge.

Several organic priority pollutants present at very low levels in influent also were concentrated in the sludge. They included acrylonitrile, (not detected in the influent vs. 41 μg/l in the combined sludge), dichlorobromomethane (0–1 μg/l vs. 74 μg/ l), and 3,4-benzofluoranthene (not detected vs. 43 μg/l).

SUMMARY

The data presented for priority pollutants in POTWs suggest several conclusions regarding the fate of toxic pollutants in POTWs.

1. Secondary treatment removes toxic metals better than primary treatment, although primary treatment does provide some significant removal.
2. In terms of organic priority pollutant removal, the preliminary findings presented indicate that aside from standard physical and biological removal mechanisms, atmospheric stripping of refractory volatile organics appears to be a significant phenomenon. Aromatic volatiles, which are resistant to biodegradation, were removed from the treatment system, but not concentrated in the sludge. It appears, therefore, that these materials were air stripped.
3. Polynuclear aromatics (PNA), which are also biologically difficult to remove, exhibited a different fate in the two POTWs studied. PNAs are less volatile than the other aromatics on the priority pollutant list, yet these materials were removed. However, the PNAs were concentrated in sludges. This seems to indicate that for these less volatile refractory materials, air stripping is not an important removal mechanism.

REFERENCES

1. Komolrit, K., and A. F. Gaudy, Jr. Biochemical response of continuous flow activated sludge processes to qualitative shock loadings. *Journal of the Water Pollution Control Federation,* vol. 38, no. 1, p. 85, January 1966.
2. Genetelli, E. J. *Biological Waste Treatment.* 35 pp. New Brunswick, New Jersey: Department of Environmental Sciences, College of Agriculture and Environmental Science, Rutgers University.
3. *Treatability of Oil and Grease Discharged to Publicly Owned Treatment Works.* USEPA, no. 440/1-75/066, April 1975.
4. Beckman, W. J., and R. J. Avendt. *Correlation of Advanced Wastewater Treatment and Ground Water Recharge.* Prepared for U.S. Environmental Protection Agency, Office of Water Program Operations.
5. Wheatland, A. B., et al. Pilot plant experiments on the effects of some

constituents of industrial waste waters on sewage treatment. *Water Pollution Control,* vol. 70, p. 626, 1971.

6. Escher, E. D., and A. J. Kicinski. Pretreatment requirements for industrial wastes discharged to municipal treatment systems. ASCE–EED Specialty Conference on Environmental Engineering Research, Development and Design, Pennsylvania State University, July 1974.

7. Beychok, M. R. *Aqueous Wastes from Petroleum and Petrochemical Plants.* New York: John Wiley & Sons, 1967.

8. *The Impact of Oily Materials on Activated Sludge Systems.* USEPA, NTIS no. PB 212-422, EPA no. 12050 DSH, March 1971.

9. *Hexane Extractable Materials and Problems at Municipal Treatment Plants.* Metropolitan Sanitary District of Greater Chicago, Department of Research and Development, report no. 75-9, May 1975.

10. *Federal Guidelines—State and Local Pretreatment Programs.* USEPA report no. EPA 430/9-76-017a, MCD 43, January 1977.

11. Pohland, F. G., and S. J. Kang. Anaerobic processes. *Journal of the Water Pollution Control Federation,* vol. 43, no. 6, p. 1129, June 1971.

12. Dague, R. R., et al. Digestion fundamentals applied to digester recovery—Two case studies. *Journal of the Water Pollution Control Federation,* vol. 42, no. 9, p. 1666, September 1970.

13. Goss, T. A. The effects of heavy metals and toxic organics on activated sludge. Masters Thesis, University of Pittsburgh, 1969.

14. Rudolfs, W., et al. Review of literature on toxic materials affecting sewage treatment processes, streams, and BOD determinations. *Sewage and Industrial Wastes,* vol. 22, no. 9, p. 1157, September 1950.

15. *A Handbook on the Effects of Toxic and Hazardous Materials On Secondary Biological Treatment Processes: A Literature Review.* Rockville, Maryland: Environmental Quality Systems, Inc., prepared for the Allegheny County Sanitary Authority and the EPA, September 1973. (Unpublished).

16. McKee, J. E., and H. W. Wolf. *Water Quality Criteria. 2nd ed.* The Resources Agency of California, State Water Resources Control Board, Publication no. 3-A, 548 pp., 1963.

17. *Environmental Effect of Photoprocessing Chemicals,* vol. II. 324 pp. Harrison, N.Y.: National Association of Photographic Manufacturers, Inc., 1974.

18. Banerji, S. K., et al. Effect of boron on aerobic biological waste treatment. Proceedings of the 23rd Industrial Waste Conference, Purdue University, p. 956, 1968.

19. Banerji, S. K., and P. R. Parikh. Effect of boron on anaerobic digestion. Proceedings of the 4th Mid-Atlantic Industrial Waste Conference, 1970.

20. Mosey, F. E. The toxicity of cadmium to anaerobic digestion: Its modification by inorganic anions. *Water Pollution Control,* vol. 70, p. 584, 1971.

21. Reid, G. W., et al. Effects of metallic ions on biological waste treatment processes. *Water and Sewage Works,* vol. 115, no. 7, p. 320, July 1968.

22. *Interaction of Heavy Metals and Biological Sewage Treatment Processes.*

201 pp. U.S. Department of Health, Education and Welfare, Environmental Health Series, Water Supply and Pollution Control, Publ. no. 999-WP-22, May 1965.

23. Moore, W. A., et al. Effects of chromium on the activated sludge process. *Journal of the Water Pollution Control Federation,* vol. 33, no. 1, p. 54, January 1961. Also published in the Proceedings of the 15th Industrial Waste Conference, Purdue University, p. 158, 1960.

24. English, J. N., et al. Slug of chromic acid passes through a municipal treatment plant. Proceedings of the 19th Industrial Waste Conference, Purdue University, p. 493, 1964.

25. Bailey, D. A., et al. The influence of trivalent chromium on the biological treatment of domestic sewage. *Water Pollution Control,* vol. 69, no. 2, p. 100, 1970.

26. Ludzack, F. J., and M. B. Ettinger. Industrial wastes—Chemical structures resistant to aerobic biochemical stabilization. *Journal of the Water Pollution Control Federation,* vol. 32, no. 11, p. 1173, November 1960.

27. Loveless, J. E., and H. A. Painter. The influence of metal ion concentrations and pH value on the growth of a *Nitrosomonas* strain isolated from activated sludge. *Journal of General Microbiology,* vol. 52, 1968.

28. Effects of copper and lead bearing wastes on the purification of sewage. *Water and Sewage Works,* vol. 93, no. 1, p. 30, January 1946.

29. Jackson, S., and V. M. Brown, Effect of toxic wastes on treatment processes and watercourses. *Water Pollution Control,* vol. 69, p. 292, 1970.

30. McDermott, G. N., et al. Effects of copper on aerobic biological sewage treatment. *Journal of the Water Pollution Control Federation,* vol. 35, no. 2, p. 227, February 1963.

31. Barth, E. F., et al. Summary report on the effects of heavy metals on the biological treatment processes. *Journal of the Water Pollution Control Federation,* vol. 37, no. 1, p. 86, January 1965.

32. Ghosh, M., and P. Zugger. Toxic effects of mercury on the activated sludge process. *Journal of the Water Pollution Control Federation,* vol. 45, no. 3, p. 424, March 1973.

33. Masselli, J. W., et al. *Controlling the Effects of Industrial Wastes on Sewage Treatment.* 62 pp. Technical Report prepared for the New England Interstate Water Pollution Control Commission by Wesleyan University, June 1970.

34. Directo, L. S. and E. Moulton. Some effects of copper on the activated sludge process. Proceedings of the 17th Industrial Waste Conference, Purdue University, p. 95, 1962.

35. Coburn, S. Limits for toxic wastes in sewage treatment. *Sewage Works Journal,* vol. 21, no. 3, p. 522, 1949.

36. McDermott, G. N., et al. Copper and anaerobic sludge digestion. *Journal of the Water Pollution Control Federation,* vol. 35, no. 5, p. 655, May 1963.

37. Masselli, J. W., et al. The effects of industrial wastes on sewage treatment. Report prepared by New England Interstate Water Pollution Control Commission, June 1965.

38. McCarty, P. L. Anaerobic waste treatment fundamentals; Part III, Toxic materials and their control. *Journal of Public Works,* November 1964.

39. Pfeffer, J. T. and J. E. White. The role of iron in anaerobic digestion. Proceedings of the 19th Industrial Waste Conference, Purdue University, p. 887, 1964.

40. *Treatability of Oil and Grease Discharged to Publicly Owned Treatment Works.* USEPA, no. 440/575/066, Pretreatment Requirements for Oil and Grease, April 1975.

41. Kugelman, I. J., and P. L. McCarty. Cation toxicity and stimulation in anaerobic waste treatment. II. Daily feed studies. Proceedings of the 19th Industrial Waste Conference, Purdue University, p. 667, 1964. Also presented in the *Journal of the Water Pollution Control Federation,* vol. 37, p. 97, 1965.

42. Nemerow, N. L. *Theories and Practices of Industrial Waste Treatment,* Reading, Mass: Addison-Wesley Publishing Company, Inc., 1963.

43. Brinsko, G. A. Annual report—Control of toxic and hazardous material spills in municipalities. Allegheny County Sanitary Authority, November 4, 1974.

44. Zugger, P. D. and M. M. Ghosh. The effect of mercury on the activated sludge process. Proceedings of the 27th Industrial Waste Conference, Purdue University, p. 792, 1972.

45. Lingle, J. W. and E. R. Hermann. Mercury in anaerobic sludge digestion. *Journal of the Water Pollution Control Federation,* vol. 47, no. 3, p. 466, March 1975.

46. Rudolfs, W., and H. R. Amberg, White water treatment. *Sewage and Industrial Wastes,* vol. 24, no. 10, p. 1278, October 1952.

47. Barth, E. F., et al. Field survey of four municipal wastewater treatment plants receiving metallic wastes. *Journal of the Water Pollution Control Federation,* vol. 37, no. 8, p. 1101, August 1965.

48. Kugelman, I. J., and K. K. Chin. Toxicity, synergism, and antagonism in anaerobic waste treatment processes. *Advanced Chemistry,* series 105, vol. 55, p. 55, 1971.

49. McDermott, G. N., et al. Nickel in relation to activated sludge and anaerobic digestion processes. *Journal of the Water Pollution Control Federation,* vol. 37, no. 2, p. 163, February 1965.

50. Mancy, K. H., and D. A. Okun. The effects of surface active agents on aeration. *Journal of the Water Pollution Control Federation,* vol. 37, no. 2, p. 212, February 1965.

51. Ghosh, S. Anaerobic processes—Literature review. *Journal of the Water Pollution Control Federation,* vol. 44, no. 6, p. 948, June 1972.

52. Lawrence, A. W., et al. The effects of sulfides on anaerobic treatment. Proceedings of the 19th Industrial Waste Conference, Purdue University, p. 343, 1964.

53. McDermott, G. N., et al. Zinc in relation to activated sludge and anaerobic digestion processes. Proceedings of the 17th Industrial Waste Conference, Purdue University, p. 461, 1962.

54. Nay, M. W., Jr., et al. Biological treatability of trinitrotoluene manufac-

turing wastewater. *Journal of the Water Pollution Control Federation,* vol. 46, no. 3, p. 485, March 1971.

55. Brown, P., and P. R. Andrew. Some effects of zinc on the performance of laboratory scale activated sludge units. *Water Pollution Control,* vol. 71, no. 5, pp. 549–554, 1972.

56. Stafford, D. A. The effect of phenols and heterocyclic bases on nitrification in activated sludges. *Journal of Applied Bacteriology,* vol. 37, p. 75, 1974.

57. Stickley, D. P. The effect of chloroform in sewage on the production of gas from laboratory digesters. *Water Pollution Control,* vol. 69, p. 585, 1970.

58. Sykes, R., and E. J. Kirsch. Accumulation of methanogenic substrates in CCl_4 inhibited anaerobic sewage sludge digester cultures. *Water Research,* vol. 6, p. 41, 1972.

59. Bischofsberger, W. Combined treatment of chemical wastes and domestic sewage in Germany. Proceedings of the 24th Industrial Waste Conference, Purdue University, p. 920, 1969.

60. Swanwick, J. D., and M. Foulkes. Inhibition of anaerobic digestion of sewage sludge by chlorinated hydrocarbons. *Water Pollution Control,* vol. 70, p. 58, 1971.

61. Potos, C. Effects of EDTA on wastewater treatment. *Journal of the Water Pollution Control Federation,* vol. 37, no. 9, p. 1247, September 1965.

62. Manganelli, R. M. Effects of synthetic detergents on activated sludge. Proceedings of the 4th Industrial Waste Conference, Purdue University, p. 611, 1948.

63. *Fate of Benzidine in the Aquatic Environment: A Scoping Study.* USEPA Contract no. 68-2226, January 1974.

64. Radhakrishnan, I., and A. K. Sinha Ray. Activated sludge studies with phenol bacteria. *Journal of the Water Pollution Control Federation,* vol. 46, no. 10, p. 2393, October 1974.

65. Winely, C. L., and C. L. Clemente. Effects of pesticides on nitrite oxidation by *Nitrobacteragilis. Applied Microbiology,* vol. 19, no. 2, p. 214, February 1970.

66. Bennett, E. R., and D. W. Ryckman. The effect of ABS shock loadings on the activated sludge process. Proceedings of the 16th Industrial Waste Conference, Purdue University, p. 52, 1961.

67. McClelland, N. I. and K. H. Mancy. The effect of surface active agents on substrate utilization in an experimental activated sludge system. Proceedings of the 24th Industrial Waste Conference, Purdue University, p. 1361, 1969.

68. Lank, J. C., Jr., and A. T. Wallace. Effect of acrylonitrile on anaerobic digestion of domestic sludge. Proceedings of the 25th Industrial Waste Conference, Purdue University, p. 518, 1970.

69. *Environmental Effect of Photoprocessing Chemicals.* Vol. 1. Harrison, New York: National Association of Photographic Manufacturers, Inc., 1974.

70. Carter, J. L., and R. McKinney. Effects of iron on activated sludge treatment. *Journal of the Environmental Engineering Division, ASCE,* vol. 99, no. EE2, p. 135. April 1973.
71. *Study of selected pollutant parameters in POTW (draft).* Prepared for USEPA by Sverdrup & Paral and Associates, Inc., February 1977.
72. Feiler, H. D., Vernick, A. S., and Storch, P. J. Fate of priority pollutants in POTW's. Eighth National Conference on Municipal Sludge Management, Miami Beach, Fla., March 21, 1977.

5
Sludge Disposal Considerations

INTRODUCTION

The U.S. Environmental Protection Agency estimates that, in this country, about 35 million metric tons of hazardous wastes are generated annually by approximately 270,000 industrial plants and other facilities. About 60% of the total generation is sludge or liquid waste (1). While implementation of pretreatment programs will reduce pollutant concentrations in POTW influent and consequently in POTW sludge, "new" sludge is generated at the industrial sites. This sludge contains toxic compounds which previously would have entered the POTW and are subject to regulations controlling sludge handling and disposal.

Viable methods of disposal for sludges resulting from pretreatment activities exist. Two common sludge disposal methods are conversion processes and land disposal. Conversion processes include incineration, pyrolysis, and composting. Landspreading and landfilling are viable land disposal alternatives (2). For industrial sludges containing high concentrations of hazardous substances, reduction, solidification, encapsulation, and fixation enhance the feasibility of the land disposal of these wastes (3,4). The basic reference guides for sludge disposal practices are the *Process Design Manual: Municipal Sludge Landfills* (2) and the *Process Design Manual for Sludge Treatment and Disposal* (5).

Most often, sludges from pretreatment operations are disposed of in sanitary landfills and by incineration. Some sludges are sent to chemical recovery facilities for processing, but the majority of the sludges are disposed of in sanitary landfill sites. Because of the strict regulations governing the disposal of these sludges in landfills, this method of disposal has become very expensive. The most pressing problem with landfill disposal is the lack of available approved and/or permitted disposal sites. Most local communities have restricted the construction or operation of such facilities within their jurisdiction because of local public opposition (6).

In Germany, the concept of segregated landfills for the separation of wastes by type to allow for future recovery is being put into practice. This method would also eliminate environmental risks due to the codisposal of wastes which could be potentially hazardous in combination. For example, sludge containing high heavy metal concentrations would be segregated from organic waste sites, since the combination of the two could result in a highly toxic leachate.

As an alternative to land disposal, there are a few high temperature incinerators in operation that can burn these sludges satisfactorily. However, this is an expense as a result of stringent air emission regulations (7).

Legislation relating to industrial sludge generation and management ensures that environmental hazards due to handling and disposal are minimized. The major pieces of legislation affecting the generation and management of sludges resulting from industrial pretreatment are outlined in this chapter. They are: Section 405 of the Clean Water Act; the Resource Conservation and Recovery Act; and section 307 of the Clean Water Act.

SECTION 405 OF THE CLEAN WATER ACT— DISPOSAL OF SEWAGE SLUDGE

Section 405 addresses municipal sludge management practices. Under this section, EPA is required to develop and publish regulations providing guidelines for the disposal and the utilization of sludge for various purposes. EPA is also required to:

1. Identify alternatives for utilization or disposal of municipal sludges;
2. Identify concentrations of various pollutants which interfere with each alternative means of sludge incineration.
3. Issue regulations governing issuance of NPDES permits for municipal sludge disposal.

Under section 405, the determination as to the manner of sludge use or disposal is a local decision. However, once a sludge management option is selected, it is unlawful for the POTW to violate federal guidelines established under section 405 for that use or disposal option. The applicable guideline from section 405 will be incorporated as conditions on the municipal NPDES permit (8).

The general pretreatment regulations provide a direct enforcement mechanism for ensuring that municipal sludge meets the requirements of section 405 of the Clean Water Act (CWA). If, for example, a POTW with an average flow in excess of 5 mgd received pollutants from industrial users which prevented the POTW from complying with the requirements of section 405 of the Act, that POTW would be required to develop a pretreatment program. The general pretreatment regulation also gives POTWs the authority to develop specific limitations in addition to federal limitations for industrial users for the purpose of ensuring that the POTW's sludge meets section 405 criteria. The regulation sets up a mechanism for regulating pollutants which "interfere" with the operation of the POTW and prevent compliance by the POTW with the requirements of section 405. In addition, the regulation provides general authority for NPDES states and/or EPA to bring enforcement actions directly against industrial users contributing pollutants which prevent a POTW from complying with section 405 of the Act. The general pretreatment regulation also interfaces with section 405 in the granting of removal allowances. The regulation provides that a POTW may not receive a removal allowance if such an allowance will result in the POTW being out of compliance with sludge use or disposal criteria, guidelines, or regulations developed under section 405 of the Act. The purpose of this provision is to ensure that any additional amount of incompatible pollutants allowed into the POTW as a result of modified pretreatment standards do not contaminate the sludge or otherwise interfere with use or disposal in a manner which harms the environment (8).

THE RESOURCE CONSERVATION AND RECOVERY ACT

The Resource Conservation and Recovery Act (RCRA) of 1976 (PL 94-580), passed by congress on September 30, 1976, is the first comprehensive federal regulation of solid waste (9). It amends the Solid Waste Disposal Act of 1965, which provided for a national research and development program for improved methods of disposal and resource recovery and for a program of technical and financial assistance to state and local governments (10). RCRA provides EPA with the ability to control and regulate all solid wastes, sludges, and hazardous residues from the "cradle to the grave". Figure 5–1 depicts EPA's cradle-to-grave approach in handling hazardous and toxic wastes (11).

The two basic objectives of RCRA are (12):

1. Protection of public health and environment.
2. Conservation of natural resources.

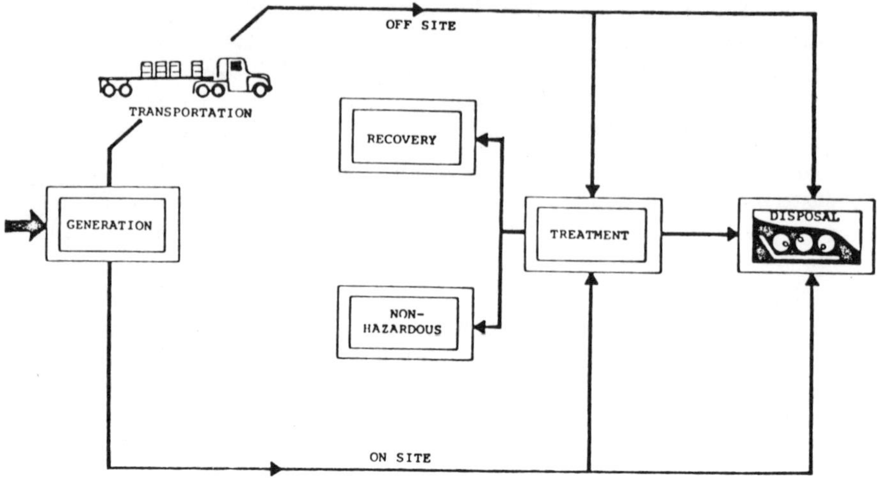

Figure 5-1 / EPA'S "cradle-to-grave" approach in handling hazardous and toxic wastes.

RCRA provides three major programs to help achieve these goals (12):

1. The establishment of a hazardous-waste control program which is to be administered by the states or by EPA in cases where the states choose not to do so;
2. The development of a land disposal regulatory program in each state;
3. The initiation and support of resource conservation programs by state and local governments to conserve resources and reduce the amount of solid waste requiring land disposal.

Hazardous-waste guidelines and regulations on identification and listing, generation, transportation, and treatment, storage, and disposal were proposed December 18, 1978 and included criteria for governing the management of hazardous wastes from generation, through treatment and storage, to ultimate disposal (13, 14, 15).

Of major concern regarding pretreatment sludge disposal are the regulations for hazardous waste management (subtitle C) and land disposal of waste that is not hazardous (subtitle D).

Seven sets of regulations and guidelines have been proposed and/or are being developed by EPA under subtitle C of the RCRA. Subtitle C provides for the following (11, 12):

Section 3001—An identification of hazardous wastes according to specified criteria. Under these proposed criteria, solid waste is considered hazardous if it is ignitable, corrosive, reactive, or toxic. In addition, lists of hazardous wastes, hazardous-waste-generating processes, and sources are set forth.

Section 3002—Standards applicable to hazardous waste generators. Regulations are set forth pertaining to the following: accurate record keeping of hazardous waste data; labeling and use of containers for storage, transport or disposal; furnishing of chemical composition information; reporting of quantities and disposition of wastes generated during a particular time period; and the use of a manifest system or tracking document to ensure the movement of hazardous wastes from the generator's premises to an approved and permitted disposal facility.

Section 3003—Standards applicable to hazardous waste transporters. These regulations would ensure the following: accurate record keeping; transporting of properly labeled wastes only; compliance with the manifest system; transporting to a permitted disposal site only; and regulation consistency with the Hazardous Materials Transportation Act and DOT regulations.

Section 3004—Standards applicable to hazardous waste treatment, storage, and disposal facilities. This section sets performance standards for the following: record keeping; reporting, monitoring and inspection, and compliance with manifest system; location, design, and construction of the facilities; contingency plans to minimize unanticipated damage; and maintenance and operation of the facilities, ownership, continuity of operation, training of personnel, and financial responsibility.

Section 3005—Permits for treatment, storage, and disposal facilities. These regulations stipulate permit requirements, application requirements, permit issuance, permit revocation for noncompliance, and interim status for specified facilities.

Section 3006—Authorization of state programs. This section includes: federal guidelines for state programs to be published; interim authorization for state programs; and specified exceptions where state programs will not be authorized.

Section 3010—Guideline pertaining to preliminary notification by generators, transporters, and treatment, storage, and disposal facilities and effective dates of regulations.

These subtitle sections and the *Federal Register* dates are summarized in Table 5-1 (16).

Subtitle D, pertaining to wastes that are not classified as hazardous, provides for the following (11, 12):

Section 4001—A description of subtitle objectives. These include assisting in developing and encouraging environmentally sound methods for solid waste disposal, maximizing resource utilization, and encouraging resource conservation.

Section 4004—Criteria for sanitary landfill. Any disposal facility not meeting those criteria will be considered an open dump.

Table 5-1 / Summary of proposed federal regulations.

Subtitle section	Title of regulation	Federal register
3001	Identification and listing of hazardous waste	December 1978
3002	Standards applicable to generators of hazardous waste	December 1978
3003	Standards applicable to transporters of hazardous waste	April 1978
3004	Standards applicable to hazardous waste facilities	December 1978
3005	Permits for treatment, storage or disposal of hazardous waste	June 1979
3006	Guidelines for development of state hazardous waste programs	June 1979
3010	Notification system	July 1978

Section 4005—Upgrading of open dumps. This section requires that any facility listed as an open dump must be closed or upgraded within five years.

Of relevance to both subtitles are the definitions set forth in section 1004 of subtitle A (general provisions), especially those of "solid waste," "hazardous waste," and "sludge." *Solid waste* is defined as "any garbage, refuse, sludge from a waste treatment plant, water supply treatment plant or air pollution control facility, and other discarded material, including: solid, liquid, semisolid or contained gaseous material resulting from industrial, commercial, mining and agricultural operations, and from community activities, but . . . not [including] solid or dissolved material in domestic sewage, or solid or dissolved materials in irrigation return flows or industrial discharges that are point sources subject to permits under section 402 of the Federal Water Pollution Control Act, as amended, or source, special nuclear or byproduct material as defined by the Atomic Energy Act of 1954, as amended" (12).

Hazardous waste is a subclass of solid waste. By definition, hazardous waste is "a solid waste or combination of solid wastes, which because of its quantity, concentration or physical, chemical or infectious character may: (a) cause, or significantly contribute to, an increase in mortality or an increase in serious irreversible, or incapacitating reversible, illness; or (b) pose a substantial present or potential hazard to human health or the environment when improperly treated, stored, transported, or disposed of, or otherwise managed" (12).

Sludge is defined as "any solid, semisolid or liquid waste generated from a municipal, commercial or industrial wastewater treatment plant, water supply treatment plant or air pollution control facility, or any other such waste having similar characteristics and effects" (12).

Also of relevance to subtitles C and D are the definitions set forth for "disposal," "manifest," "open dump," "sanitary landfill," "storage," and "treatment."

Also of importance is the relationship between RCRA and other laws as specified in section 1006 (application and integration with other acts), including the Federal Water Pollution Control Act; the Safe Drinking Water Act; the Marine Protection, Research, and Sanctuaries Act; and the Atomic Energy Act. In practice, this section means, for example, that an industrial point source discharge covered by an NPDES permit would not also require an RCRA permit for that discharge, but disposal of sludge from that water treatment facility would be covered by RCRA (12).

At this point in time, it appears that the approach EPA will take in reference to CWA and RCRA in resolving the municipal sludge situation is as follows (11, 17):

1. Municipal sludges will not be covered under subtitle C, (hazardous waste management) of RCRA regulations.
2. Section 405 regulations will be as restrictive as or more restrictive than the requirements developed under sections 3004 and 4004 of RCRA.

Significant penalties exist for failure to comply with RCRA. A generator of wastes that complies with RCRA regulations could be liable to third parties in the event of improper disposal by a disposal contractor. The liability issue is a complex one and focuses on strict liability, negligence, nuisance, and independent contractor relationships (12).

SECTION 307 OF THE CLEAN WATER ACT— TOXIC AND PRETREATMENT EFFLUENT STANDARDS

Section 307 contains several different requirements relating to the identification and control of toxic pollutants. Under the authority of section 307, prohibited discharge standards and categorical pretreatment standards are established and ideally result in lower concentrations of toxic pollutants into POTW sludge. Although extensive information on pretreatment exists and a positive effect on POTW sludge seems certain, the following factors make it difficult to accurately estimate the reduction in the pollutant concentrations in the sludge (7):

1. The effectiveness of pretreatment technologies in removing toxic compounds present in the waste can range from 10% to 90%.
2. Nonindustrial resources of pollutants can contribute significantly to the total toxic pollutants entering a POTW and thereby reduce the effectiveness of a pretreatment program.
3. The effectiveness of implementation of EPA's pretreatment strategy on the local level can vary.

ENVIRONMENTAL FACTORS AND PUBLIC CONCERN

Legislation regarding pretreatment has been implemented to ensure maximum environmental protection. This protection falls into several categories. A primary goal is to ensure the efficiency and continuity of treatment plant operations. Plant upsets or interference, which can prove to be detrimental to receiving water quality, can be minimized by limiting contributions of incompatible pollutants or excessive discharge of compatible pollutants. Similarly, by controlling hazardous discharges that are explosive or corrosive, the protection of the physical POTW facilities can be assured. Even if plant upsets were not a problem, the contribution of incompatible hazardous substances alone would be an environmental impetus for the development of pretreatment programs. If the POTW is not designed to treat these incompatibles, an adverse impact on the receiving waters can result from their passing through a POTW. Additionally, removal of incompatible pollutants at the POTW can degrade sludge quality, thereby limiting sludge disposal options (18).

While one of the goals of pretreatment is to reduce pollutant concentrations in the POTW influent (and consequently in the POTW sludge), thereby ensuring greater treatment efficiency, enhanced feasibility of land treatment of sludges, and fewer overall environmental hazards, some environmental risks are associated with sludges generated from pretreatment facilities. The public's support of federal efforts to regulate hazardous industrial sludge may be weakened by a lack of confidence that such wastes can be managed by local agencies. This lack of confidence is most acute when plans to develop sludge disposal sites are initiated. The "anywhere-but-near-me" syndrome is even more severe when the site would be used for disposal of hazardous sludge (1).

The public's hostile reaction is due largely to a misunderstanding of the waste management industry and its modern, professional disposal practices; an inherent fear of anything described as hazardous; and the extraordinary focus by the media on illegal and ignorant waste disposal practices that have surfaced, often with disastrous consequences. The infamous Love Canal incident, in which increased miscarriages and birth defects resulted from contamination of an area by a neglected landfill for toxic chemical industrial wastes, has increased the public's distaste for the dumping of hazardous materials (19). The public's substantial fears of *indiscriminate* dumping relate to contaminated drinking water supplies and recreational facilities, the increasing prospect of cancer and other diseases, and even the possibility of death (1). What is not gaining as much attention is the fact that most of the publicized indiscriminate dumping horror stories involve dump sites that are no longer used; that in many of these cases, federal, state, and local efforts have successfully cleaned up these abandoned land disposal sites; and other handling and disposal of sludges resulting from pretreatment, although they may contain hazardous substances, is *strictly* regulated and *does not* involve indiscriminate procedures.

The public's opposition to sludge disposal is a major factor inhibiting the expansion of disposal facilities. What the public does not realize is that there is already a shortage of suitable disposal facilities and that the problem will only become more critical as the quantity of hazardous wastes increase (1). Although the public's attitude toward the disposal of sludges is not entirely justified, it is understandable.

Changing that attitude will take more than proposed rules and regulations. It will require massive public reeducation as well. Public information and education programs provided for under section 208 of the Clean Water Act can be successful channels for public involvement and reeducating.

REFERENCES

1. Hazardous sludge management rules: Criteria, procedures and facility standards. *Sludge Magazine,* vol. 2, no. 1, p. 12, 1979.
2. *Process Design Manual—Municipal Sludge Landfills.* USEPA, Technology Transfer, EPA-625/1-78-010, 1978.
3. Landreth, R. E., and J. L. Mahloch. Chemical fixation of wastes. EPA Technology Transfer Sludge Treatment and Disposal Seminar, 1977.
4. Pojasek, R. B. Solid-waste disposal: Solidification. *Chemical Engineering,* p. 141, August 13, 1979.
5. *Process Design Manual for Sludge Treatment and Disposal.* USEPA, Technology Transfer, EPA-625/1-74-006, 1974.
6. Whitebloom, S. W., R. Lanyon, and C. Lue-Hing. Impact on sludge production and energy consumption for industrial pretreatment systems resulting from USEPA pretreatment regulations. The Metropolitan Sanitary District of Greater Chicago, 1979.
7. Denit, J. D., and C. J. Dysinger. Impact of pretreatment legislation on sludge. Proceedings of the Eighth National Conference on Municipal

Sludge Management: Impact of Industrial Toxic Materials on POTW Sludge, Miami Beach, Florida, March 1979.

8. Hutzel, N. Impact of the general pretreatment regulation on municipal sludge practices. Proceedings of the Eighth National Conference on Municipal Sludge Management: Impact of Industrial Toxic Materials on POTW Sludge, Miami Beach, Florida, March 1979.

9. Tough cleanup rules set for chemical dumps. *Chemical Engineering,* vol. 85, no. 28, p. 56, 1979.

10. Bryson, J. E. Solid waste and resource recovery. In: *Federal Environmental Law,* ed. E. L. Dolgin and T. G. P. Guilbert. St. Paul, Minnesota: West Publishing Co., 1974.

11. Local pretreatment program requirements and guidance. Springfield, Virginia: Environmental Technology Consultants, Inc., June 1979. (Unpublished.)

12. Glaubinger, R. S. A guide to the Resource Conservation and Recovery Act. *Chemical Engineering,* p. 79, January 29, 1979.

13. Weddle, B., deputy director, state programs and resource recovery division, USEPA. Personal communication, 1979.

14. Lazar, E., manager, sludge program land disposal division, USEPA. Personal communication, 1979.

15. Corson, A., chief, waste characterization branch, hazardous and industrial waste division, USEPA. Personal communication, 1979.

16. Edwards, E. A hopeful sign. *EPA Journal,* vol. 5, no. 2, p. 9, 1979.

17. Bastian, R., environmental scientist, municipal technology branch, USEPA. Personal communication, 1979.

18. Vernick, A. S., H. D. Feiler, and P. D. Lanik. Elements of an industrial waste control program. In: *Pretreatment of Industrial Wastes—Joint Municipal and Industrial Seminar.* USEPA, Technology Transfer, Seminar Handout, 1978.

19. Hartenstein, W., Love Canal task force on-site coordinator, New York state department of environmental conservation. Personal communication, 1979.

Industry's Role in Pretreatment

INTRODUCTION

Simply stated, industry's primary role in pretreatment is to conform to the local, state, and federal regulations applicable to the specific industrial facility. Although this function on the surface would appear to be relatively straightforward, it nevertheless becomes increasingly complex when one considers the number, magnitude, and requirements of the regulations involved. Additionally, the technical, economic, institutional, and administrative information required by a specific industrial plant to allow it to comply with pretreatment regulations is voluminous and involves a number of detailed factors.

This chapter provides the basic information which will permit an industry to:

—Define the nature of its wastewater discharge by means of an in-plant survey;
—Consider in-plant measures to minimize such discharges through best management practices;
—Work with the POTW involved;
—Establish removal credits through the POTW to reduce categorical pretreatment standards;
—Analyze financial considerations and options;
—Develop the most effective approach to reduce costs;
—Select the necessary pretreatment processes to treat specific wastewater discharges;
—Evaluate applicable sludge treatment and disposal options.

IN-PLANT WASTEWATER SURVEY

A comprehensive survey of plant waste streams constitutes the basic tool in developing any industrial wastewater management program. Essentially, the purpose of this type of a survey is to provide all of the vital data relative to wastewater systems within the plant. As such, it represents the starting point or baseline for subsequent wastewater discharge control plans.

A thorough survey serves to define the current status of wastewater conditions in the plant, thereby permitting comparison with applicable pretreatment requirements. In this regard, the wastewater survey has a twofold purpose:

1. It furnishes the data necessary to achieve conformance with federal, state and local pretreatment regulations and standards.
2. It provides for the development of a design basis for wastewater control and treatment facilities.

The establishment of accurate, current wastewater data for industries discharging to a POTW has assumed new significance with the promulgation of the federal pretreatment regulations in June 1978. Previously, indirect dischargers had to provide information on the characteristics of their discharges to their POTW as part of the NPDES permit program. That requirement frequently could be satisfied with a single sampling of the plant effluent. Complying with the new pretreatment regulations requires a more in-depth analysis, and an examination of the facility's wastewater generation, collection, and treatment systems by means of a complete survey of plant wastewaters.

As previously indicated, a survey of plant wastewaters is also necessary to develop a design basis for wastewater control and treatment facilities. The general and categorical federal pretreatment regulations will undoubtedly have the effect of requiring many more plants to install pretreatment facilities. Frequently a comprehensive wastewater survey will provide the data necessary to judge the merits of centralized treatment as opposed to treatment of individual or groups of waste streams. At the same time, the survey will furnish needed information on wastewater characteristics, enabling comparison with the applicable pretreatment standards. In this way, the degree of treatment necessary prior to discharge can be accurately established.

In addition to regulatory considerations and definition of treatment requirements, a survey of plant wastewaters can help to accomplish several other significant goals. A complete definition of the flow and pollutant characteristics of each wastewater stream generated in the plant can frequently lead to significant conservation measures. A survey can identify areas where recycling and reuse of wastewater streams is possible, and often uncover process changes that might easily be instituted to reduce water usage and/or pollutant loading in individual waste streams. Frequently, significant savings in water usage, with a corresponding cost reduction, can be achieved by the objective examination of all aspects of wastewater generation in the plant.

An additional incentive for a complete survey of plant wastewaters is the general desire to achieve segregation of incompatible wastewater streams. Most industrial facilities have separate sewer systems to handle stormwater, process wastes, and sanitary sewage within the plant. However, in many older facilities cross-connections of these systems are common, resulting in contamination of streams that could be reused, contamination of clean storm runoff, excessive hydraulic loading, and general inefficiency in the collection and treatment of wastewater.

In summary, the purpose of a comprehensive survey of plant wastewaters is to provide the information necessary to:

1. Achieve compliance with applicable federal, state and local pretreatment regulations and standards;
2. Develop a design basis for wastewater control and treatment requirements;
3. Conserve resources through reduction of water usage and pollutant loading wherever possible;
4. Segregate incompatible wastewater streams and sewer systems.

There are two basic elements in any plant wastewater survey:

1. Definition of the physical characteristics of the plant's sewer systems;
2. Development of individual wastewater stream profiles.

Generally, completion of many of the tasks necessary to define the physical systems is required before information on the flow characteristics and pollutant loading of individual waste streams can be obtained. Consequently, the two major components of the survey are usually handled consecutively rather than simultaneously, although some overlap is inevitable. Figure 6–1 provides a graphic illustration of the various components of a wastewater survey (6).

Physical Characteristics

The most common starting point of a plant survey is in the drawing file, where the extent of available information can be determined (4). All layout and schematic drawings which relate to the plant sewer systems and to water usage within the plant, should be compiled and carefully previewed. If overall sewer maps and schematic flow diagrams of the drainage systems are not available, they must be prepared. If these drawings do exist, they must be updated. The end product of this phase of the survey should be a complete

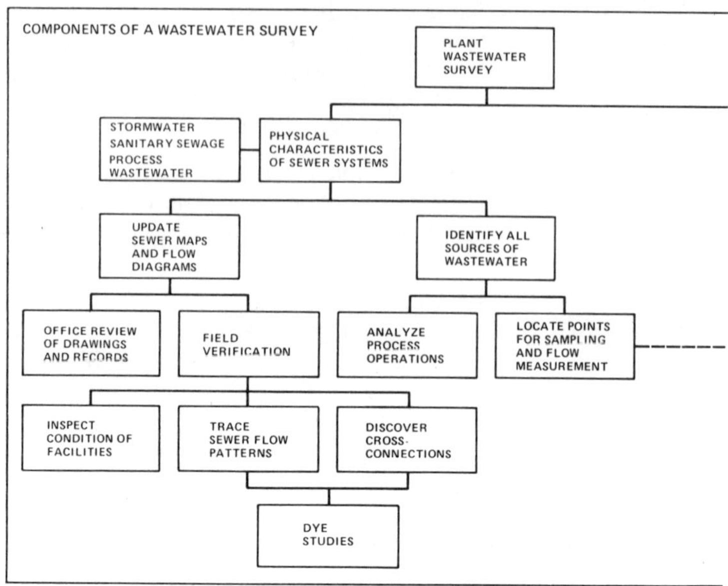

Figure 6-1 / Components of a wastewater survey.

set of up-to-date sewer maps and a comprehensive flow diagram of the plant's wastewater systems.

In gathering data, attention should be given to available information on the sanitary sewage system and stormwater collection and disposal facilities, as well as the process drainage system. In many plants, concern with the sanitary sewage system will focus on segregating sanitary wastes from other plant wastes, and ensuring that there are no cross-connections between the separate systems. With regard to stormwater runoff, generally this phase of the survey should confirm that clean runoff is truly uncontaminated and is segregated prior to discharge, and that all contaminated runoff is directed to process sewers. Conditions will obviously vary from plant to plant, but in most instances the greatest amount of time and effort will be expended in gathering data to fully define the process sewer system.

In reviewing record drawings and other available data, emphasis should be placed on the water supply system to delineate and document all processes using water within the plant. Generally, process operations should be studied to determine how water is used, and, in turn, discharged from each operation. Some processes actually produce water by synthesis as a by-product of a chemical reaction, while others consume water in various ways. By this type of review the actual mechanism of wastewater production within the plant can be established, which will then serve as a basis for the additional investigative work which must be performed.

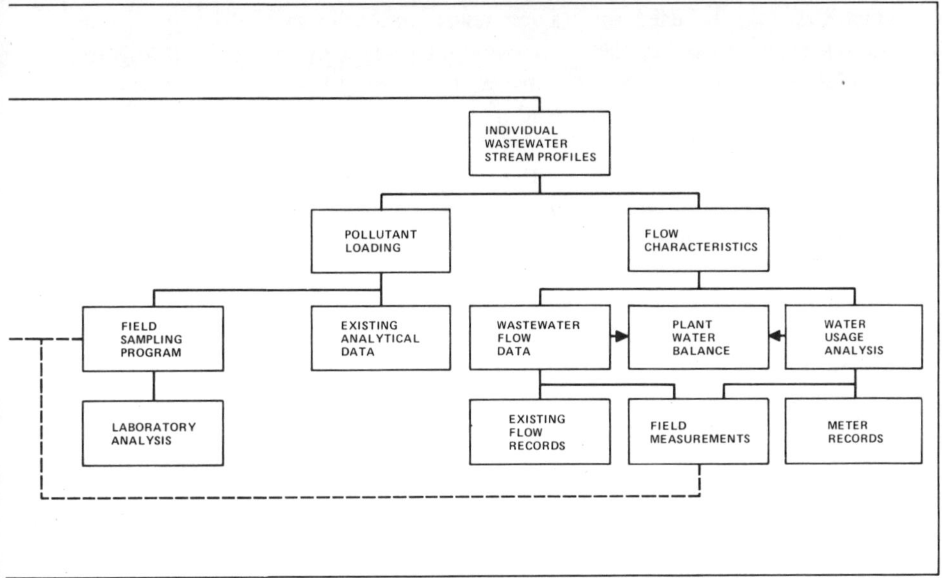

Upon completion of the office review of drawings and other information, the next step involves field verification of data gathered thus far, and exploration of areas of uncertainty. This phase of the survey is extremely important and requires a dedication to detail, as well as a determination to uncover all necessary information. The overall objective of this portion of the field work is to visually confirm information shown on the drawings and to fill any data gaps by first hand observation. Specific objectives include:

1. Complete definition of physical characteristics to permit updating of all sewer maps and flow diagrams;
2. Identification of all sources of waste within the plant;
3. Location of sampling points for individual and composite waste streams;
4. Discovery of cross-connections between storm, process, and/or sanitary sewer systems;
5. Familiarization with existing collection and disposal systems for subsequent work.

In performing the field work, all wastewater streams should be traced from one end to the other, that is, from the waste source to the point of ultimate disposition. This involves inspection of all manholes, not only to verify location and condition, but primarily to confirm or determine flow direction in the sewer. Frequently, dye studies are necessary to establish beyond question the pattern of sewer flow within each system. Several chemicals may be used for tracing sewer lines with dye, although sodium fluorescein is probably the most common. It is a hygroscopic, orange-red powder which is freely soluble in water, perceptible to the naked eye at a concentration of 0.02 mg/l, and has no toxic effect on fish (5). These properties, combined with its brilliant yellowish-green fluorescence, make it a relatively simple task to trace the flow in sewers with sodium fluorescein.

The field investigation should include interviews with key operating and maintenance personnel throughout the plant. Frequently, their knowledge of details and field changes not shown on the drawings can be invaluable in establishing sewer location and flow patterns. In old plants, long-time employees often are the best source of information concerning sewer lines. In any event, data obtained in this fashion should be verified by visual inspection and dye studies, if necessary.

Completion of the first phase of the survey will fully define the physical characteristics of the plant's sewer systems, and set the stage for the development of individual waste stream profiles. It will also produce the data required to update sewer maps and flow diagrams or produce new drawings in the drafting room, at the same time as the remainder of the survey proceeds. It is generally advantageous to make all drawing revisions as soon as possible after completion of the field investigations. In this way, the information is still current, and questions may be resolved in the field conveniently as the ensuing portion of the survey is being completed.

Waste Stream Profiles

The establishment of individual waste stream profiles, which characterize the flow and pollutant loading from all wastewater sources, is perhaps the most important portion of any plant wastewater survey. It provides the basis for all subsequent evaluations and judgments inherent in a wastewater management plan. Consequently, considerable care should be exercised in this monitoring activity to ensure that all data obtained is accurate and representative of actual plant conditions. The two components of a stream profile, flow and pollutant loading, are generally and most conveniently determined at the same time in the field. However, since the considerations involved in each determination are significantly different, they have been addressed independently in the following sections.

Flow Characteristics

The accurate determination of individual waste stream flows involves a broad overall evaluation of water usage within the plant. This can be accomplished most easily through a water budget analysis or water balance, which accounts for all individual water demands, and compares water usage to the field measurement of individual wastewater streams.

Generally a good starting point in the water balance evaluation is with a review of plant records on water usage and consumption. If water is obtained from a municipality or private water company, then the monthly or quarterly water bills provide a basis for establishing total plant consumption. If water is obtained directly from wells or a surface source of supply, generally it is metered as it enters the plant. These records should be carefully examined for a period of several years, not only to establish an average daily usage, but to determine trends, seasonal variations, or other significant factors as well.

Many industrial facilities meter the water supplied to major process units or other significant users within the plant. Records of this type can be extremely helpful in determining the pattern of water distribution within the facility. Drawings showing the water supply system should be examined and field checked to establish an accurate picture of all water usage. Most often, field measurements and estimates will be necessary to determine individual flow values. Many methods are available to estimate flows where records and flow instruments are unobtainable. Frequently, equipment ratings, tank gauges, process analysis, industry standards, and visual observations may be utilized to estimate specific flows. Attention should be paid to periods of operation, so that total consumption values can be determined for intermittent water users. Care must be exercised to ensure that values established for each water usage are representative of conditions that usually prevail within the plant.

In most instances, actual field measurement of incoming flow is desirable to either determine a specific usage value or confirm a previous estimate. Since water is supplied in pressure conduits, water meters are the most applicable for measuring fresh water flow. In larger lines, orifices, nozzles, and venturi meters can be utilized, while nozzles and orifices can also serve to compute the flow from freely discharging pipes. Frequently, water supply lines will have to be broken to install a measuring device. Although this procedure may be troublesome and time consuming, it is nevertheless invaluable in obtaining accurate water usage data.

When this aspect of the water balance has been completed, a summary such as the one shown in Table 6-1 should be prepared. The summary must be concise, but still provide sufficient detail to fully document each entry. The total plant water usage obtained in this manner should be in reasonably good agreement (preferably less than 10%) with the consumption value for the entire plant obtained from meter records, water bills, etc. If a significant disparity results from this comparison, further field work is in order to refine the individual usage figures.

Table 6-1 / Plant water usage.

Water use	Average consumption, gpd	Comments
A. Domestic (sanitary sewage)	8,600	Estimated 20 gal/capita/day per accepted plumbing practices.
B. Utilities		
1. Cooling tower makeup	28,400	Average of water meter records for past two years.
2. Steam boiler makeup	18,000	Average of direct measurements made over a two-week period.
3. Water treatment	800	Estimated from equipment manufacturer's data for backwash; 80 gpm for 10 minutes once per day.
C. Unit A		
1. Water used in product	60,000	Average of production records for past two years.
2. Laboratory	4,800	Ten sinks running continuously during an 8 hour day; estimated at 1 gpm by bucket and stopwatch.
3. Pump and compressor cooling	20,200	Direct single measurement—14 gpm—24 hours/day.
4. Floor wash	600	Estimated from hose nozzle size, water pressure, and time of operation.
5. Reactor wash	500	Reactors 1, 2, 3; five times/year 4, 5—twice yearly. Estimated average daily flow.
D. Unit B		
1. Vacuum seal water	7,200	Rated at 10 gpm while operating; average operation—12 hours/day.
2. Compressor cooling	9,800	Two compressors: (1) Unit 1 rated at 1 gal/100 ft^3. (2) Unit 2 rated at 3 gpm when operating.
3. Venturi scrubbers	16,800	Single measurement of 12.5 gpm. Plant operators estimated an average of 13 to 15 gpm. Assumed 14 gpm for 20 hours/day.
4. Floor wash	500	Same as unit A.
5. Vessel rinse water	1,200	Internal spray rinse—1000 gpd; Hand rinse—200 gpd; per operating personnel.
E. Pilot plant	4,000	Pump cooling measured at 15 gpm. *Seal water measured at 7 gpm.*
Total	181,400	

The next task in the water budget analysis is the determination of flow values for all individual waste streams entering the sewer systems. This includes those flows normally directed to the sanitary and storm water sewers as well as those directed to the process sewer system. In most cases, flow data on individual waste streams will not be available, so that field measurements and estimates will be necessary. In most plants, particularly those discharging to publicly owned treatment works, the entire plant effluent is monitored for flow, as in Figure 6-2, prior to discharge. In those cases, the records should be reviewed to determine representative average and peak values, which will serve to check the cumulative total obtained from the individual waste stream flow determinations.

This task requires attention to detail and ingenuity as do estimating and measuring individual water supply flows. Prior to initiating field work, a

Figure 6-2 / Monitored effluent.

program should be developed which delineates the location, frequency, and type of flow measuring device to be utilized for each determination. The program should be based on information gathered during the earlier field work, and should include methods to be used for estimating the flow of specific waste streams where actual measurements are not practical. The format for the program should be established so that entries can be made when the data is obtained, and a summary of wastewater flow information such as Table 6-2 is automatically produced.

Actual wastewater flow determinations in the field involve consideration of a number of factors concerning the process itself. Variability of flow caused by process conditions is extremely important. Consequently, details concerning batch vs. continuous operation, seasonal or diurnal process variations, and the duration and nature of the process, require judgments regarding the type, frequency, and extent of the monitoring program. In order to avoid extensive piping modifications for the installation of monitoring equipment, innovative techniques should be utilized wherever possible. A number of references and technical articles are available which deal with this subject in great detail. Specifically recommended is the *Handbook for Monitoring Industrial Wastewater* (1) available through EPA technology transfer, in addition to the two references on monitoring cited in Chapter 3. One of those references includes information on the performance of portable automatic flow measuring devices. There are several devices of this type that are commercially available which can provide reasonably accurate flow measurement data, and which would be useful in a plant wastewater survey.

Table 6-3 (6) provides a summary of the types of methods that are available for use in determining wastewater flows within the plant. Generally, the choice of a specific method will depend primarily upon accessibility, equipment, piping configuration, and hydraulic characteristics (i.e., flow from freely discharging pipes vs. flow in open channels, etc.). For flow from the open end of a pipe, as in Figure 6-3, the bucket-and-stopwatch method can be a convenient procedure for relatively small flows. It relies upon the use of a container of known volume and notation of the time required to fill it. Pipes which are flowing full when freely discharging may be monitored by the use of orifices and nozzles in a manner similar to that used with pressure conduits. The flow from pipes which are partly full when discharging can be estimated by either the open pipe or the California pipe method, or measured directly by use of an open flow nozzle (1).

The most common method for measuring the flow in sewers or open channels is by the use of weirs or flumes, which create an obstruction to the free flow of liquid. A direct measurement of water level at a specified distance upstream of the weir or flume in turn establishes the flow, which is proportional to the depth of liquid (3). The use of these devices in a plant survey may require temporary installations, where drainage lines are cut and

Figure 6-3 / Open pipe discharge.

directed to a measuring flume or weir specially rigged for the purpose. Flow recording instruments are readily connected to these devices and can be extremely helpful in establishing flow values and patterns. Other techniques for flow determination in sewers include the use of current meters and pitot tubes, the measurement of liquid depth and velocity, and the chemical dilution method (2). In addition, related observations, such as measuring level changes in tanks, pumping rates and duration, etc., can frequently be employed to estimate the flow in sewer lines.

In addition to determining the flow for individual waste streams, the total plant effluent should be monitored if this is not routinely done. The average value established in the field or from past records for the plant effluent should be compared to the total obtained from summarizing the individual wastewater flows (Table 6-2). As with the water supply figures,

Table 6-2 / Wastewater flows.

Wastewater source	Process sewer system		Sanitary sewer system		Stormwater sewer system		Discharge frequency	Comments
	Average flow, gpd	Peak flow, gpd	Average flow, gpd	Peak flow, gpd	Average flow, gpd	Peak flow, gpd		
A. Domestic (sanitary sewage)			8,200	12,400			8 hr/day	Average and peak values of flow monitoring records for past two years.
B. Utilities								
1. Cooling tower blowdown	5,400	8,200					continuous	Average and peak values of direct readings by temporary flume over a two week period.
2. Boiler blowdown	2,000	—					continuous	Estimated by open pipe method.
3. Water treatment backwash	900	—					10 minutes once/day	Single measurement by bucket and stopwatch
C. Unit A								
1. Laboratory	4,800	—					continuous for 8 hrs.	Same measurement as plant water usage (Table 6-1)
2. Pump and compressor cooling	20,200	—						Same measurement as plant water usage (Table 6-1)
3. Floor wash	600	—					once/day	Daily 10 minute hose down.
4. Reactor wash	500						19 washes/yr	Estimated average daily flow.
5. Roof and area drainage					intermittent			Volume, frequency and duration dependent upon weather conditions.

Source						Frequency	Remarks
6. Steam condensate	3,000	—		6,500	—	continuous	Condensate is only partially returned to the boiler, with the remaining volume being discharged at eighteen individual points. Each flow was estimated by bucket and stopwatch.
7. Reactor upset and spillage							Emergency condition of a random nature.
D. Unit B							
1. Vacuum seal water	7,200	—				3 times/day	Same estimate as plant water usage (Table 6-1)
2. Compressor cooling	11,600	—				continuous	Determined by measuring depth of flow in the sewer.
3. Venturi scrubbers	16,800	—				20 hr/day	Same estimate as plant water usage (Table 6-1).
4. Floor wash	500	—				once/day	Daily 10 minute hose down.
5. Vessel rinse water	1,200	—				3 tanks/day	Same estimate as plant water usage (Table 6-1).
6. Roof and area drainage				intermittent			Volume, frequency and duration dependent upon weather conditions.
7. Steam condensate	4,200	—		3,700	—	continuous	Condensate is only partially returned to the boiler, with the remaining volume being discharged at fourteen individual points. Each flow was estimated by bucket and stopwatch.
E. Pilot plant	4,200	—				intermittent	Determined from observation of sump pump operation; pump rated at 30 gpm and operates for an average of 6 minutes per hour.
System totals	83,100		8,200	10,200			
Total to sewer			101,500				

213

Table 6-3 / Flow measurement techniques.

Method	Application	Comments
Venturi meter	A	Useful where high pressure recovery is necessary.
Orifice meter	A, B	Relatively inexpensive; easy to install; large head loss.
Flow nozzle	A, B	Greater head loss and less cost than a Venturi meter; less head loss and more cost than an orifice.
Magnetic flow meter	A	Good for fluids containing suspended solids; no greater head loss than for equivalent length of pipe.
Pitot tube	A, D	Economical for large pipes; can be used to measure the velocity flow in an open channel.
Rotameter	A	Simple, inexpensive device for measuring relatively low flow rates of clear, clean liquids.
Open-end pipe method	B, C	Approximate inexpensive procedure for estimating flow from open pipes by measuring two dimensions of the stream as it is freely discharging into the air.
California pipe method	C	Requires a horizontal freely discharging pipe of known diameter and measurement of the distance from the top of the pipe to the water surface.
Open flow nozzle	C	Two common types are Kennison nozzle and parabolic flume; useful for accurate and continuous measurement.
Weirs	D	Commonly used; easy to install; inexpensive; V-notch, Cipolleti or rectangular design; sediment can accumulate behind a weir and affect accuracy.

Flumes	D	More accurate than a weir; self-cleaning; Parshall or Palmer–Bowlus design; minimum upstream effects; subject to flooding.
Depth of flow measurement	D	Utilization of the Manning formula to estimate flow by measuring the depth, when the slope and coefficient of roughness are known.
Continuity equation ($Q = AV$)	D	Involves determination of velocity by a current meter or timing a float or tracer dye between two points, and measuring the depth of flow to determine the wet cross-sectional area.
Bucket and stopwatch	B, C	Estimating procedure using a container of known volume and noting the time required to fill; good for small flows where physically convenient.
Pumping rates	A, B, C	Flow determination by recording the time of pumping and pump capacity at the discharge pressure using manufacturer's pump curves.
Tank level change	A, B, C	Useful for batch operations to determine waste flow by monitoring the change in level with time in a tank of known volume.
Dilution method	A, D	Addition of dye, salt, or radium to the liquid stream with corresponding measurement of concentration to determine flow; monitoring apparatus is expensive.

Applications:
A. Pressure pipes.
B. Freely discharging pipes flowing full.
C. Freely discharging pipes flowing partly full.
D. Open channel flow.

agreement within 10% should be considered acceptable. If a larger disparity results, judgment is necessary to select the areas requiring reexamination in the field.

When all field measurements and estimates have been completed, the last step in the water budget analysis is the comparison of water supply and wastewater flow values as illustrated in Table 6–4 (6). In this procedure, allowances must be made for all water consumed or lost in the plant in order to effect a true balance. Typical usage within a plant resulting in consumption would include water used in the product, evaporation and windage losses from a cooling tower, makeup to a steam boiler, lawn sprinkling and/or irrigation, and leaks in the water supply system. In older facilities, infiltration into sewer lines should be taken into account in reconciling sewer flows with water supply figures. The final water balance should provide an accurate summary and accounting of all water used in the plant, which in turn should serve as a key element in fulfilling the goals of the plant wastewater survey.

Pollutant Loading

The second major component, in addition to flow characteristics, required to develop waste stream profiles is pollutant loading. This term refers to the quantity of each pollutant parameter present in a specific wastewater stream. In establishing representative pollutant loading values for individual waste streams, previous records or sampling results should first be examined. If such data are available in sufficient detail, a field monitoring program may only be necessary to update and verify the existing information. However, if data are unavailable or insufficient, a full sampling and analysis program is indicated. The sampling program must be integrated with the flow determinations as stated previously, and it must give careful consideration to the location, type, frequency, and duration of sampling to be done in the field. A working knowledge of each process is necessary to ensure that all samples obtained are representative of actual conditions, taking into account such factors as the effect of changes in production on wastewater characteristics, diurnal variations, seasonal effects, etc.

An important consideration is to establish which pollutant parameters to analyze for. In this regard, a helpful approach in complex operations is to monitor the plant effluent for several days, and to have these samples analyzed for the full spectrum of pollutant parameters. This procedure will confirm the presence, or document the absence, of each pollutant parameter. Pollutants known to be present in the plant effluent can then be checked in the individual waste stream analyses.

Field sampling can be accomplished either manually or through the use of automatic devices. The three distinct types of sampling procedures described in Chapter 3, namely, grab sampling, simple composite sampling, and flow-proportioned composite sampling, may be utilized. Automatic sampling devices can also be helpful in a wastewater survey, and as indicated previously, these devices vary over a wide range of cost, applicability and reliability. The considerations relevant to the potential for errors in field sampling and the procedures to be utilized to avoid such errors, which were discussed in Chapter 3, are also applicable to the in-plant survey.

The analytical aspects of the sampling program also involve a number of detailed considerations. Many plants have their own laboratory facilities or have direct access to a corporate analytical laboratory. In these cases, a decision must be made as to whether to utilize the company lab, or whether to employ an independent laboratory for analysis of the samples. Many factors should influence this decision, including the capability of the company laboratory in terms of personnel and equipment to perform the required analyses, and the proximity, capability and cost of an independent laboratory. Another important consideration is the question of objectivity, in that it may be preferable to utilize an independent lab for all data that will be reported to a regulatory agency.

If the option of an independent lab is chosen, the actual selection of a firm is extremely important. Quality control within the lab must be exercised to ensure the accuracy and reproducibility of the results. A technique for checking the accuracy of laboratory results is to submit preanalyzed samples for analysis by the lab. Such spiked samples are made available to interested parties by the regional analytical quality control coordinator in each of the ten EPA regional offices.

All sample collection, preservation, and analysis must be done in accordance with the precise and recognized techniques established for the analysis of wastewaters, as discussed previously. Analytical results must be carefully reviewed and evaluated to ensure that the data is accurate and truly representative. Variability in results can usually be attributed either to abnormal circumstances in the plant (e.g., temporary changes in the process, upset conditions, poor housekeeping or spills, etc.) or to errors in sampling, analysis, or calculation. Questionable results should be brought to the attention of the laboratory for reexamination and confirmation. If necessary, additional samples should be taken at this time in order to check previous values. When this has been accomplished and there is reasonable certainty that the data is accurate, it should be summarized in a clear and systematic manner.

Analysis of Data

The final step in the development of individual waste stream profiles is the application of statistical analysis to the data. Statistical correlation will permit a proper choice of average values and will also provide a correct

Table 6-4 / Water balance.

Water usage	Average volume supplied, gpd	Average wastewater flow, gpd			Consumptive use	Comments
		Process sewer	Sanitary sewer	Storm sewer		
A. Domestic	8,600		8,200			
B. Utilities						
1. Cooling tower	28,400	5,400			24,700	Calculated evaporation and windage loss based upon circulating rate and location.
2. Steam boiler	18,000	2,000				
a. Unreturned condensate		7,200		10,200		
3. Water treatment	800	900				
4. Misc. leaks in water supply system					5,000	Allowance based on estimates in the literature for system in this age category.
C. Unit A						
1. Water used in product	60,000				60,000	
2. Laboratory	4,800					
3. Pump and compressor cooling	20,200	20,200				
4. Floor wash	600	600				
5. Reactor wash	500	500				

D. Unit B
   1. Vacuum seal water
   2. Compressor cooling
   3. Venturi scrubbers
   4. Floor wash
   5. Vessel rinse water

E. Pilot plant

D. Unit B					
1. Vacuum seal water	7,200	7,200			
2. Compressor cooling	9,800	11,600			
3. Venturi scrubbers	16,800	16,800			
4. Floor wash	500	500			
5. Vessel rinse water	1,200	1,200			
E. Pilot plant	4,000	4,200			
System totals	181,400	83,100	8,200	10,200	89,700
Allowance for infiltration	—	4,000	400	—	—
Adjusted system totals	181,400	79,100	7,800	10,200	89,700
Water balance	181,400		186,800		
Discrepancy		5,400 = 3% ±			
		Acceptable			

method for determining the range of values for a specific parameter. Generally, average values and variability as expressed by the standard deviation will be of most interest in developing a wastewater management plan. This information should be sufficient to provide the basis for judgments regarding the control of individual wastewater streams, process changes, recycling, segregation of wastewaters, and the evaluation of alternative programs. However, consideration of peak values and probability determinations such as the 50% and 90% values will undoubtedly be necessary for the design of wastewater treatment facilities. One of the many available books or technical articles on statistics should be consulted for this type of data analysis.

BEST MANAGEMENT PRACTICES— UTILIZING THE SURVEY RESULTS

The term *best management practices* (BMPs) was created as a result of specific requirements of the Clean Water Act of 1977. Section 304(e) of the act established new authority for the Environmental Protection Agency to protect surface waters from toxic and hazardous pollutants through best management practices to be imposed by means of the NPDES permit system. Subsequent EPA regulations defined best management practices to mean "methods, measures, or practices to prevent or reduce the contribution of pollutants to waters of the United States. BMPs include, but are not limited to, treatment requirements, operating and maintenance procedures, schedules of activities, prohibitions of activities, and other management practices to control plant site runoff, spillage or leaks, sludge or waste disposal, and drainage from raw material storage."

From a regulatory standpoint, best management practices apply only to industrial facilities discharging directly to navigable waters. BMPs will be imposed through the NPDES permits, and BMP plans will have to be developed by effected dischargers. Although the BMP regulations do not legally cover indirect dischargers, the BMP concept nevertheless is fully applicable to industrial plants that discharge to POTWs.

This concept implies a thorough knowledge and understanding of the generation of wastewater within the plant, and control over the collection, handling, treatment and discharge of all wastewaters. In the broadest sense, best management practices for industries within a POTW system can be interpreted as full utilization of the information developed by the in-plant wastewater survey.

In general terms, the survey results should lead to a plan for wastewater management within the plant, and the development of a continuing program for monitoring and control of plant wastewaters, as illustrated by Figure 6-4 (6). As indicated at the outset, an in-depth plant survey frequently will bring to light information not previously realized regarding water usage and loss of product or raw materials. This data can in turn be used for conservation purposes, which can result in considerable cost savings. The initial emphasis should be placed on a thorough appraisal of each waste stream to investigate the various possibilities available to reduce water usage and pollutant loading. Often, process changes can be instituted with a minimum of disruption, which will significantly reduce the quantity of water used or the amount of pollutants discharged into the waste stream. In many instances, water recycling schemes will become evident once the waste streams with the major pollutant loadings are isolated. Finally, the elimination of cross-connections between the sewer systems, resulting in wastewater segregation, can frequently effect sizable cost savings.

If wastewater treatment is necessary, the results of the survey will provide all of the data required to proceed in a systematic manner. The feasibility and advantages of treating individual waste streams, or groups of waste streams with compatible pollutants, can be explored and evaluated before considering centralized treatment. In many cases, this type of approach will be advantageous, in that treatment would be performed prior to dilution in the sewer by other waste streams, and considerably smaller volumes would be involved in the treatment process. In any event, the data gathered during the plant survey will provide the basis for the evaluation of alternative treatment programs. Comparison of wastewater data with applicable pretreatment regulations should serve to provide a basis for a preliminary conceptual design of facilities. In most cases, a detailed design for pretreatment of industrial wastewater should not be undertaken without a pilot program to firmly establish all design parameters.

Once the wastewater situation within the plant has been fully defined and a wastewater management plan has been formulated, required information can be furnished to regulatory agencies and a cost estimate for implementation of the plan can be developed. At this point, plant management should be briefed on the current situation and recommended course of action. This is necessary not only to obtain authorization of funds required to proceed, but also to keep management informed on the environmental aspects of plant operation. These condsiderations in turn impact on corporate image, public relations, relationships with governmental agencies, and legal requirements for pretreatment of wastewater discharges.

The final aspect of utilizing the results of the wastewater survey is the establishment of a continuing program for wastewater monitoring and control. The extensive effort expended in completing an in-depth survey should not be regarded as a one-time endeavor. Rather, it should be viewed as the beginning of a continuing plan to control wastewaters within the plant. Such a program requires emphasis from management to provide the necessary funds, and to create an awareness and orientation in all employees toward

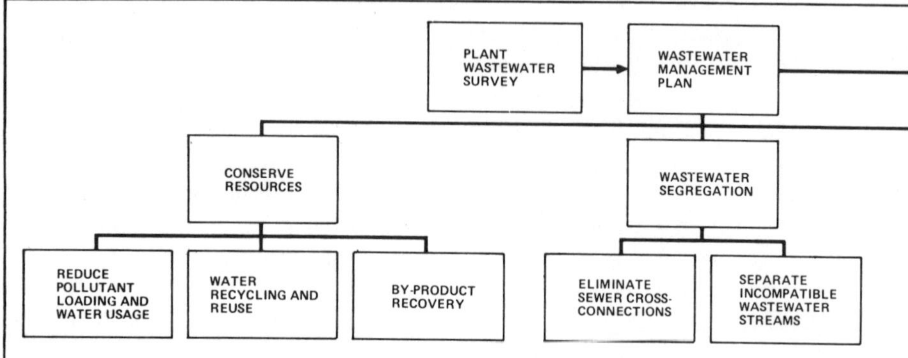

Figure 6-4 / Best management practices: wastewater survey utilization.

environmental considerations. This can be accomplished by stressing the fact that waste monitoring and waste treatment systems are in reality an integral part of plant operations. Production and operating personnel should be encouraged to consider the variability of their waste load as a factor in evaluating the overall performance of their unit.

In the final analysis, wastewater control depends not only upon the operation of treatment facilities and the performance of monitoring equipment, but equally upon the awareness, attitude and concern of all plant personnel.

WORKING WITH MUNICIPALITIES

In the past, it was not uncommon for industry to view governmental agencies at all levels in the environmental field as adversaries. Many industrial organizations were uninformed regarding the mandate, direction, and specific requirements of environmental regulations, and frequently waited to be told what to do by regulatory agencies having jurisdiction over their facilities. However, the passage of Public Law 92-500 in October 1972 constituted a turning point in many respects with regard to the awareness and attitude on the part of industry concerning environmental protection and pollution control.

The sweeping mandate of that legislation and subsequent laws and court decisions have made it obligatory for industry to become totally familiar with the laws and regulations applicable to it. In addition, in spite of the many legal actions that resulted from the effluent guidelines program, industry in general has realized that it must work with regulatory agencies to avoid excesses and attempt to achieve workable regulations that it can live with. Nowhere is this cooperative spirit more important than in the relation-

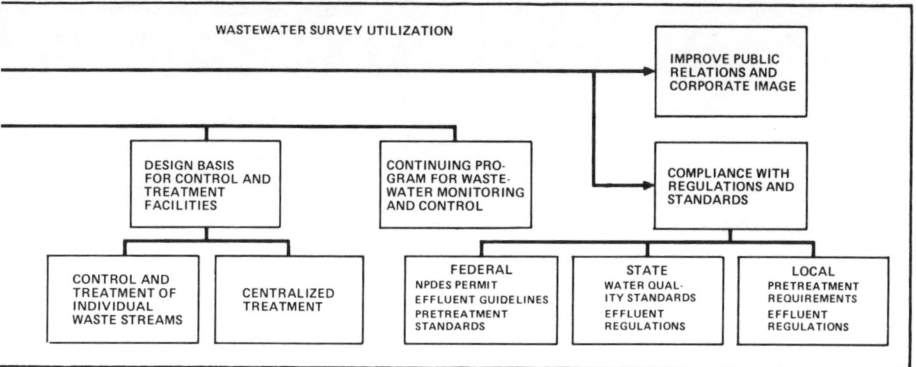

ship between a specific industrial plant and the POTW into which it discharges its wastewater.

Need for a Working Relationship

Chapter 3 provided information relating to the public relations aspect of POTW personnel working with industry. It was emphasized that POTWs embarking upon a pretreatment program must, from their point of view, establish a working relationship and spirit of mutual trust with the industries in the system. It is equally important from the industry point of view that such a relationship be developed.

The POTW will ultimately be the enforcement agency for federal categorical pretreatment standards, in addition to enforcing the general pretreatment standards and local ordinances. The need for an industrial facility to have a working relationship with its enforcement agency should be self-evident. In addition, many industrial plants may qualify for a removal credit resulting from the removal or reduction of specific pollutants occurring in the POTW treatment process. Such removal credits may be used to reduce the categorical pretreatment standard applicable to the plant discharge. Removal credits must be applied for by the POTW, hence the obvious need for industry and POTW personnel to work together. The subject of removal credits is discussed in detail in Chapter 2.

Establishing a Working Relationship

Where industry is selecting personnel for wastewater management responsibility, a knowledge of the background of the counterpart individual in the POTW is desirable. The relationship between industry and the POTW is, like any other business relationship, founded on the relationships between people. To make it succeed, the people need to relate favorably toward each other (7).

The relationship can be cultivated and developed through several mechanisms. It was suggested in Chapter 3 that the POTW sponsor informal coordinating committees, consisting of representatives of the affected industries and the public agency. Industry's willingness to participate in such arrangements, and actually taking the initiative by volunteering facilities, secretarial services, etc., can be extremely helpful to POTWs with limited resources.

Many industries have training seminars in wastewater management, sampling, analysis, etc., either set up for the waste control personnel at a single plant, or for all the waste control personnel of a multiplant company. Some seminars are operated by trade associations, or, in some cases, by an educational agency. Such industry-oriented training programs are highly recommended. Whenever possible, the POTW personnel should be invited to participate. Such joint seminars can help to develop the type of good working relationship between industry and POTW personnel that is mutually beneficial.

Another avenue for interface and communication between industry and POTW representatives is through professional organizations and meetings. The industry management should encourage the waste control manager to become a member of the local water pollution control association, which automatically provides him with membership in the Water Pollution Control Federation. Industry representatives should participate in sectional, state, and national meetings and volunteer for appropriate committee activities. Attendance at such technical conferences and committee work provides exposure to people from municipalities and state and federal agencies, as well as representatives of other industries with similar interests and problems.

A final vehicle for establishing a working relationship is through joint research leading to the development of a technical article or paper. Many papers on industrial wastewater technology presented at technical conferences are joint papers by technologists from industry and POTWs. Joint interest in getting answers, collecting and evaluating data, and producing new ways to solve old problems produces a durable mutual trust that is a valuable asset for both the industrial user and the POTW authority (7).

FINANCIAL CONSIDERATIONS

Financial Options

Industrial facilities discharging into publicly owned sewer systems must consider a number of options and strategies in order to minimize the economic impact of compliance with pretreatment regulations. Typically, industrial plants discharging into POTWs have been faced with the cost of user charges, surcharges, and industrial cost recovery charges as the price for use of the public sewer and treatment works. These costs were described in detail in Chapter 3, with an emphasis on the municipality's obligation to make such charges fair and equitable.

Industry's options regarding user charges, surcharges, and industrial cost recovery charges are usually limited to either leaving the POTW completely in favor of direct discharge, or providing pretreatment to minimize the impact of these charges. Such evaluations must be made on a case-by-case basis, considering the specific charges involved, the characteristics of the plant's waste stream, and the cost of pretreatment if the facility is to remain in the POTW system. If abandoning the POTW in favor of direct discharge is contemplated, then additional considerations include the availability of a stream for discharge of the plant effluent, applicable water quality regulations and effluent limitations, ease of obtaining an NPDES permit, and the cost of installing treatment facilities to satisfy regulatory requirements. These basic questions must be evaluated and resolved in order to establish an appropriate course of action for a specific industrial facility.

Financial Assistance

If pretreatment is either necessary to satisfy pretreatment regulations or selected to minimize the cost of user charges, surcharges, etc., then a number of options are available with regard to obtaining financial assistance. Each company must individually evaluate the financial assistance alternatives available, to select the ones most consistent with its specific goals and management objectives. Therefore, it is impossible to develop a universal prescription for an optimal financing strategy for pretreatment investments. Rather, the alternatives described below should be carefully analyzed to determine which programs are most applicable and consistent with the needs and goals of a particular industrial facility.

Financial assistance for pretreatment facilities encompasses federal, state, and local programs which include investment tax credits, tax exemptions on development bonds, loan subsidies, and rapid amortization procedures for pollution control equipment (9). These programs are available from the federal government through the Small Business Administration, the Environmental Protection Agency, the Economic Development Administration of the Department of Commerce, the Farmers' Home Administration, and the Internal Revenue Service. The assistance includes programs especially designed to stimulate investment in pollution control facilities, in addition to assistance programs of a general type to encourage industrial investment in capital facilities of any kind.

Internal Revenue Service—Income Tax Provisions

Businesses installing pretreatment equipment may choose between two methods of income tax treatment which provide investment incentives; an investment tax credit or rapid tax amortization.

Investment Tax Credit

Under the provisions of the Tax Reform Act of 1976, a firm is allowed to take 10% of the cost of certain capital investments as a credit against its federal tax liability. The investment tax credit is an instrument of industrial policy intended to encourage capital investment and expansion. While not specifically designed to aid firms installing pollution control equipment, investments in the purchase and installation of such equipment are among the investments eligible for the tax credit.

To receive an investment tax credit, the company chooses to depreciate the pretreatment equipment using any IRS-approved depreciation method, including straight line, sum of the year's digits, and declining balance. To qualify for the full investment credit, the facility must be depreciable, have a minimum three-year useful life, be a tangible, integral part of the enterprise's operations, and be placed in operation during the year for which the credit is sought (9). Capital investments eligible for the tax credit have the following limitations based on their expected useful life (8):

Useful Life	Percentage of Cost of Property Qualifying for Credit
Under 3 years	0
3 years or more, but less than 5 years	$33^{1}/_{3}$
5 years or more, but less than 7 years	$66^{2}/_{3}$
7 years or more	100

The amount of investment credit which may be claimed in a taxable year generally cannot exceed total tax liability or $25,000 plus 50% of any tax liability in excess of $25,000. There is also a $100,000 limitation on used property that can be taken into account as qualified investment for purposes of the investment tax credit for a taxable year.

Rapid Tax Amortization

The Tax Reform Act of 1976 reinstituted rapid tax amortization provisions which had previously been in effect from 1969 through the end of 1975 for pollution control facilities. Rapid tax amortization was originally provided by congress to reduce the resistance of private industry to the installation of pollution control equipment required to meet new regulations.

The 1976 act provides for the availability of a five-year amortization election for pollution control facilities which were placed in service after 1975, and used in connection with a plant or other property in operation before January 1, 1969. The facility must abate or control water or atmospheric pollution or contamination by removing, altering, disposing, or storing pollutants, contaminants, wastes, or heat, and must have been certified by the state and federal pollution control authorities as being in conformity with applicable state and federal regulations. In the case of a treatment facility used in connection with a plant not in operation before 1969, but in operation before 1976, only a portion of the investment may be rapidly amortized. Thus the rapid amortization provision is clearly intended to aid relatively older manufacturing operations (9).

In addition, the installation of eligible equipment cannot lead to a significant increase in output or capacity, a significant extension of useful life, a significant reduction in operating costs, or a significant alteration in the nature of a manufacturing or production process or facility. Significance is defined for the purposes of this provision as a change of greater than 5% (8).

If the facilities qualify as outlined above, the taxpayer is allowed to recover costs over a 60-month period. However, this 60-month amortization deduction is limited to facilities with a useful life of no more than 15 years, or that fraction of the basis of a facility with a longer life allocable to the first 15 years. In addition, the act provides for the availability of one-half of the investment tax credit (5%) for eligible pollution control facilities. The combination of rapid amortization with 50% of the investment tax credit is intended to make rapid amortization more attractive to industry than previously.

Rapid amortization under the 1976 law is attractive for pollution control equipment with depreciable life longer than 11 to 12 years. For equipment with a shorter life, standard amortization yields greater tax savings (8). Also, the magnitude of the potential tax benefits provided over standard amortization practice varies as a function of discount rate and depreciable equipment life, with rapid amortization being attractive only at high discount rates (9).

Companies planning to take advantage of either the investment tax credit or rapid amortization provisions of the tax laws would be well advised to consult the Internal Revenue Service or the economic analysis division of EPA for further details.

Small Business Administration Programs

The Small Business Administration (SBA) is the sponsor of several programs applicable to the financing of pretreatment facilities. A company falls within the SBA definition of a small business if it is "independently owned and operated, is not dominant in its field of operations, does not have assets exceeding $9 million, does not have net worth in excess of $4 million, and does not have an average net income, after federal income taxes for the preceding two years, in excess of $400,000." (9). Businesses must qualify as well under SBA rules which set size standards for individual industries.

Loans

Section 7 of the Small Business Act, as amended, authorizes loans to assist small business concerns to add or to alter their equipment, facilities, or methods of operation in order to meet the requirements of the Federal Water Pollution Control Act. The loans are to provide assistance to eligible small businesses, farmers, and feedlot operators who might otherwise incur a substantial economic injury. The Environmental Protection Agency must certify that the equipment for which the loan is requested will be sufficient to bring the firm into compliance with the applicable environmental regulations (8).

Any small business firm subject to pretreatment requirements may qualify for SBA assistance. Assistance in the form of direct loans or guaranteed/insured loans is available for making additions to or alterations in equipment, facilities (including *specifically* pretreatment facilities), or methods of operation (9).

Loans are limited to $500,000, unless substantial hardship is proven. The SBA share of guaranteed loans may not exceed 90%. Direct loans to eligible firms are provided when commercial loan sources are unavailable, or can provide only part of the required financing. Loans are made for up to 30 years at interest rates significantly below the prevailing rates available for private financing (8).

To qualify for a direct loan from SBA, the applicant must demonstrate substantial injury, meaning in effect an inability to obtain conventional financing, either because of insufficient collateral, excessive interest or an excessively short repayment term. A supporting letter from a bank is necessary. Because conventional lenders are allowed to charge 10% interest for SBA guaranteed/insured loans, direct loans from the SBA are clearly preferable. In the past, virtually all loans have been direct loans from SBA, rather than guaranteed loans obtained from a bank (9).

Pollution Control Revenue Bonds

In 1976, an amendment to the Small Business Act authorized a program for SBA to guarantee the payments on tax-free municipal bonds for small businesses for the purpose of financing pollution control equipment. The statute specifically provides that the financing of the pollution control facilities be funded through industrial revenue bonds, issued by a state or municipality, on which the interest is exempt from federal income tax.

Through this program, it is hoped that small businesses will have access to capital markets similar to those available to large companies which finance pollution control facilities through tax-free industrial development bonds. A public entity, usually the state, issues the tax-exempt revenue bonds for industrial pollution control projects. The business receiving the loan from proceeds of the bond signs a contract for making payments to the public entity, and a trustee for the entity uses those funds to pay bond holders and make redemptions (9).

The primary purpose of using tax-exempt industrial revenue bond financing for pollution control facilities is to obtain the most advantageous interest rates and repayment terms possible. Tax-exempt bonds may be issued with a sinking fund provision for redemption at particular dates, or may be in serial form with varying maturities up to 25 or 30 years. The interest rate may be the same for the full issue or it may vary according to maturity and market conditions. The interest rate for a bond issue with SBA's guarantee of the underlying payments would be comparable to AAA municipal bond rates (8).

Because of the substantial risk involved in a 100%, 25-year SBA guarantee of the bonds, applicants must have been in business a minimum of 5 years and have a history of profitable operations on average over the last 5 years. Applicants must also provide evidence of the need for the pretreatment facility or equipment. The program is generally most suited to pretreatment projects costing in excess of $100,000 (9).

Economic Development Administration Loans

The Economic Development Administration (EDA) of the U.S. Department of Commerce is charged with the responsibility of stimulating the industrial growth and development of economically depressed areas of the country. EDA business development loans are made to firms in an attempt to upgrade an area economically, primarily through the creation of new and permanent jobs and higher income for local residents (8).

While pollution control considerations are not specifically related to the EDA's objectives of alleviating economic distress and increasing employment, the business development loan program can indirectly help firms to install pollution control equipment. Companies located in a low-growth area as designated by the EDA, may be eligible for a loan if there is the likelihood that the firm would be shut down and jobs lost for noncompliance with environmental regulations.

Financial assistance is directed toward those firms which are unable to arrange assistance from banks or other private lending institutions. Long-term loans up to 25 years at a specified treasury rate are available covering up to 65% of the cost for the acquisition of fixed assets: 10% of the costs must be in the form of applicant's equity; 5% will normally come from a local development corporation or state agency; the remainder must be financed by a conventional commercial lender (9).

EDA also has the authority to make direct loans and to guarantee loans up to 90%. Typically, EDA loans are large and are oriented toward large businesses not covered by SBA programs. In 1977, EDA provided 27 business loans and 20 guarantees, ranging from $260,000 to $5,200,000, with an average of $1,500,000. However, fewer than five of those were for pollution control (9).

Environmental Protection Agency Grants

Construction Grants for Municipalities.   Under the construction grant program, EPA is authorized to make grants for the construction of new municipal sewage treatment facilities. The federal funding share for these projects is 75%. The rest of the initial costs are divided among state and local governments and industrial users who hook up to the POTW.

The law requires that industrial users of municipal systems financed through federal grants pay their proportionate share of total construction, operation, maintenance, and replacement costs. Accordingly, when industrial users hook up to these facilities, they must pay:

1. A user charge based on the O & M costs incurred by the POTW in treating their discharges;
2. An industrial cost recovery charge to repay the portion of the Federal grant which is allocable to the treatment of their industrial wastes.

Although the industrial cost recovery requirement ensures that industrial users will repay their share of the federal grant, capital cost recovery may occur over a period of 30 years and there need be no interest charges. The municipal grants program may therefore provide a significant benefit to industrial users (8).

Direct Grants to Industry for Research and Development.   Section 105 of the Clean Water Act provides for direct grants to industrial facilities for research and development purposes. The grants cover the construction of permanent wastewater treatment facilities on industry property to develop and demonstrate new or improved methods of treating industrial wastewaters which have industry-wide application. Grants so authorized may not exceed 75% of the projected costs. In addition, the act also provides grants for projects which can develop and demonstrate "new or improved methods of joint treatment systems for municipal and industrial wastewaters."

The grants program is a cost-sharing type of agreement. However, in some of EPA's research and development the federal government bears the full cost, contracting with appropriate nongovernment entities for these services. Through the contract or grant device, EPA can support almost any related worthwhile project under almost any type of institutional arrangement.

Two basic conditions must be met before a grant is made. First, the project must have scientific and technical merit. Second, the project activities must usefully demonstrate a way to combat water pollution. The results must be accurate in terms of both cost and performance, and they must be made available to the public (8).

Farmers Home Administration— Business and Industrial Loans

This program was authorized by the Consolidated Farm and Rural Development Act, and is intended to assist, among others, private organizations in rural areas to obtain loans for the purpose of improving the economic and environmental climate, including pollution abatement and control. FmHA business and industrial loans may be made in any area outside the boundary of a city of 50,000 or more and its immediately adjacent urbanized areas with population density of more than 100 persons per square mile. Priority is given to applications for projects in open country, rural communities, and towns of 25,000 population and smaller (8).

FmHA assistance ordinarily is provided as a loan guarantee whereby the agency contracts to reimburse the lender for not more than 90% of principal and interest. Applicants apply for the guarantee through their private lenders. Repayment of the loan varies, but final maturity cannot exceed 30 years for land, buildings, and permanent fixtures; 15 years for machinery; or the life of the equipment, whichever is shorter.

The basic purposes of FmHA business and industrial loans include developing or financing business or industry, increasing employment, and controlling or abating pollution. Consequently, the financing of pollution control expenditures represents only a small part of the purpose of the business and industrial loan program. FmHA encourages borrowers of less than $500,000 to use the SBA, as available FmHA funds are currently inadequate (9).

State and Local Programs

Industrial Development Bonds

Industrial Development Bonds (IDBs) are sold by state and local governments to help companies obtain financing necessary to meet federal pollution control requirements. These pollution control bonds usually are sold through a state or municipal agency to underwriters who resell them to investors. Funds raised must be spent for meeting air and water pollution control requirements.

Interest on IDBs is not subject to federal income tax, even though generally that type of bond is not tax exempt. In effect, private borrowers are allowed to use the tax-free market normally available only to governmental agencies. The net effect is that the bonds are generally easier to sell. The borrowing costs to the corporation are normally superior to those rates available from private lending institutions, or from the sale of corporate bonds (9).

This type of financing has been extensively utilized by large businesses for their financing of pollution control facilities. Typically, an industrial firm planning to construct, install, or purchase pollution control or abatement facilities by this method must first persuade a local public body or state agency to issue bonds to cover costs. Essentially, the public entity issues tax-exempt revenue bonds, on which repayment is based solely on the credit of the business.

Once the bonds are issued, an interest and principal repayment schedule is established. The payments are made by a trustee for the public entity from payments received from the business in fullfillment of the contract between the business and the state or municipality. The public entity is the nominal owner of the property; the property is conveyed to the business under a lease, lease-purchase, installment sales, or similar contract (8). During the financing period, the borrower is treated as the owner for tax purposes, and may take depreciation and investment tax credits.

Tax Exemptions

Practically all states provide for corporate income tax deductions and sales tax and real property tax exemptions for industrial wastewater treatment facilities certified by the designated state agency.

Generally, state pollution control tax exemptions provide that the value of property used in a corporation's pollution abatement program can be deducted from the total value of that corporation's assessable property. For example, a company that installs pollution control equipment at a plant would be able to subtract the value of that equipment from the total value of the corporation's assessable property. However, any additional land or buildings necessary to locate or house the pollution control equipment would be classified as real property, and thus would be tax exempt only if the state has pollution control tax exemptions for real property (8).

When granting sales or use tax exemptions, states must confirm that the company involved is purchasing or installing equipment actually designed for pollution control. Equipment or machinery which normally is considered part of the industrial production process would not qualify for an exemption and would be assessed the normal sales or use taxes.

PRETREATMENT PROCESSES

The range of treatment processes available to industrial facilities discharging to POTWs encompasses the full spectrum of technology utilized in the environmental engineering field. The selection of a pretreatment process for a particular industrial plant of course depends upon the characteristics of the wastewater being discharged. In turn, the pollutant parameters present in a specific discharge result from the nature of the industrial operation being conducted at a given site. Typically, knowledge of the specific industrial processes utilized within an industrial category can be employed in a broad sense to predict the characteristics of the wastewater from a plant within a particular industry.

Table 6-5 summarizes the significant pollutant parameters present in the effluent from each of 34 major industrial categories (10). The table covers traditional pollutant parameters such as BOD, COD, suspended solids, etc. and the heavy metals. It does not include information on the priority pollutants cited in the Clean Water Act of 1977, which the Environmental Protection Agency is currently evaluating. The categorical pretreatment standards to be promulgated for 34 specific industrial categories will cover many of these pollutants.

Despite general information of this type, pretreatment of industrial wastes before discharge into a POTW system is invariably a case-by-case problem. In general, it is usually required to meet federal, state, or local pretreatment regulations to prevent the discharge of pollutants which may interfere with or pass through municipal treatment plants. Typically, despite the regulations, pretreatment decisions are dictated by the specific circumstances of each individual situation. An industrial facility discharging directly to navigable waters may decide to join a POTW system and pretreat after a technical and economic analysis because it is the most cost-effective alternative. Similarly, industries in the system may opt for pretreatment

Table 6–5 / Summary of significant pollutant parameters for major industries

Industry	BOD	TSS	TDS	COD	pH	Cyanide	Color	Oil and Grease	Phosphorus	Ammonia	Fluoride	Chrome
Dairy	X	X	X	X	X		X		X			
Grain mills	X	X	X	X	X				X			
Canned and preserved fruits and vegetables	X											
Canned and preserved seafood	X	X		X				X				
Sugar	X	X	X	X	X							
Textile	X	X	X	X	X		X	X				X
Cement												
Feedlots	X	X	X	X						X		
Metal finishing & electroplating		X			X	X			X		X	X
Organic chemicals		X	X	X				X		X		
Inorganic chemicals		X	X	X	X							X
Plastics and synthetics	X	X	X	X	X			X		X	X	X
Soap and detergents	X	X	X	X	X			X				
Fertilizer		X		X					X	X	X	
Petroleum	X							X		X		X
Iron and steel	X					X		X		X		
Nonferrous metals		X	X	X				X			X	
Phosphates		X	X	X					X		X	
Steam electric power		X		X				X	X			X
Ferroalloys		X			X	X				X		X
Leather tanning & finishing	X	X		X	X			X				X
Glass	X	X	X	X	X		X	X	X			
Asbestos	X		X	X	X							
Rubber	X	X	X	X				X				
Timber	X	X		X	X		X	X				
Pulp and paper	X	X		X			X					
Builders paper	X	X		X								
Meat	X		X	X			X	X				
Paint and ink	X	X	X	X	X		X	X				X
Auto and other laundries	X	X	X	X	X			X				X
Water supply		X	X	X			X				X	
Steam supply		X		X				X	X			X
Misc. foods and beverages	X	X		X	X			X				
Misc. chemicals	X	X	X	X	X	X		X				X

where a small investment in pretreatment facilities would result in a significant reduction in the pollutant loading and a corresponding large reduction in surcharge or user charge fees.

In all cases, as indicated above, in-plant control measures which might reduce or eliminate the need for pretreatment facilities should first be examined before embarking on a pretreatment program. In many circumstances, a thorough examination of plant operational practices, recycle alternatives, and other water conservation or reuse possibilities can significantly reduce pollutant loads.

Copper	Lead	Zinc	Cadmium	Iron	Nickel	Arsenic	Sulfide	Manganese	Mercury	Nitrate Nitrogen	Phenol	Boron	Selenium
X	X	X	X	X	X								
											X		
									X				
											X		
		X					X			/	X		
	X	X					X	X					
X	X	X	X			X							X
		X		X		X							
X		X		X							X		
				X				X			X		
							X						
											X		
	X	X											
											X		
X	X	X	X	X					X				
X	X	X	X	X	X	X							
X		X	X								X		
		X		X							X	X	X

Pretreatment considerations differ depending on whether the pollutants
to be controlled are susceptible to treatment in a POTW. For such pollutants, pretreatment would generally be employed because it is the most cost-
effective alternative, or because the POTW does not have available design
capacity to treat the wastes. Even in cases where the municipality has
available design capacity to accept an industrial discharger without pretreatment, the industry should still perform a break-even economic analysis to
determine if pretreatment is cost effective. In cases where design capacity is
not available, the municipality may elect to assign waste load allocations to

each industrial discharger, thus fixing the required pretreatment. Waste load allocation will in these cases be necessary to meet the municipality's NPDES permit requirements.

Industrial wastewaters susceptible to treatment in POTW's generally are organic in nature. Unit processes to treat such wastes can take a variety of forms, including (10):

1. Coarse solids separation;
2. Grit removal;
3. Equalization;
4. Neutralization;
5. Dissolved air flotation;
6. Sedimentation;
7. Biological treatment:
   a. Activated sludge systems;
   b. Trickling filter systems;
8. Physical-Chemical treatment:
   a. Chemical coagulation;
   b. Filtration;
   c. Activated carbon adsorption.

Most pretreatment systems should include some form of coarse solids separation, grit removal, and equalization. Equalization may be required for either flow, pollutant load, or both. It is particularly important in the prevention of excessive discharge and the imposition of shock loads on POTW systems. Similarly, neutralization is a significant unit process in pretreatment operations, to avoid possible process upset or damage to POTW facilities by highly acidic or alkaline wastes.

Dissolved air flotation and sedimentation are utilized to remove suspended solids and floatable material such as oil and grease. Dissolved air flotation is particularly useful in industries where oil and grease presents problems in the discharge of wastewater to municipal systems. Biological treatment is used in pretreatment operations to reduce BOD loading prior to discharge to a municipal sewer. The desired reduction is highly variable, and depends upon economic considerations and available capacity in the treatment facility. In many instances, a BOD reduction to 200–250 mg/l is sought in pretreatment facilities to produce an effluent which simulates domestic sewage.

Frequently, high-rate activated sludge systems or roughing filters are utilized in pretreatment systems to effect the desired BOD reduction in the most cost-effective manner. Physical-chemical treatment may be employed for suspended solids and BOD removal where technical and economic factors favor this alternative over conventional primary and biological treatment processes.

Industrial wastewaters that are not generally susceptible to treatment in a POTW system are typically inorganic in nature. Pretreatment processes for these incompatible pollutants can also take a variety of forms, including (10):

1. Coarse solids separation;
2. Grit removal;
3. Equalization;
4. Neutralization;
5. Dissolved air flotation;
6. Sedimentation;
7. Filtration;
8. Chemical precipitation and coagulation;
9. Activated carbon adsorption;
10. Chemical conversion.

The treatment of incompatible pollutants is frequently concerned with the removal of inorganic suspended and dissolved solids. The unit processes involved in inorganic suspended solids removal, such as sedimentation and dissolved air flotation, involve the same considerations as with compatible pollutants. Similarly, equalization and neutralization are equally important in treating incompatible pollutants to prevent shock loading and possible process upset or damage to POTW facilities.

The distinguishing element in treating incompatible pollutants is the removal of inorganic dissolved solids, particularly metals. The most common processes employed are chemical precipitation and coagulation and chemical conversion. In precipitation and coagulation, the dissolved pollutant reacts with the chemical agent used to form an insoluble precipitate. The precipitate is allowed to settle and is removed as a sludge. In chemical conversion, the pollutant in question is converted to another less harmful form or to another substance. Typical of this type of treatment is the reduction of hexavalent chromium to the trivalent form and the destruction of cyanide. Activated carbon adsorption may be used to remove dissolved organic pollutants which are not susceptible to treatment in POTWs.

In summary, pretreatment must be evaluated on a case-by-case basis within the context of cost effectiveness and applicable technical factors. Pretreatment considerations depend primarily upon whether the wastewater in question is compatible or incompatible with POTW systems. In either case, a variety of unit processes are available to a particular industrial facility for pretreatment of wastewater prior to discharge to a municipal sewer.

An excellent reference for information on pretreatment unit processes is the fact sheets contained in the EPA *Innovative and Alternative Technology Assessment Manual.* These two-page descriptions of wastewater treatment processes provide a convenient synopsis of the significant characteristics and relevant considerations for each specific process. The fact sheets cover virtually all of the unit processes most commonly utilized for pretreatment purposes by industries discharging to POTWs.

GENERAL REFERENCES

Girling, R. M. Practical methods for determining sewage flow for all communities. *Water and Sewage Works,* pp. 250–258, July 1969.

Innovative and Alternative Technology Assessment Manual, USEPA, Publication no. EPA 430/9-78-009, MCD 53, March 1978.

Manual of Methods for Chemical Analysis of Water and Wastes. Cincinnati, Ohio: USEPA, Methods Development and Quality Assurance Research Laboratory, National Environmental Research Center, 1974.

Methods for Collection and Analysis of Water Samples for Dissolved Minerals and Gases. Techniques of Water-Resources Investigations of the United States Geological Survey, Book 5, Chapter Al, U. S. Department of the Interior, 1970.

Miller, H. In-plant water recycling—Industry's answer to shortages and pollution. *Design News—OEM,* pp. 81–88, November 5, 1973.

Pruett, J. M. Using statistical analysis for engineering decisions. *Plant Engineering,* pp. 155–157, May 13, 1976.

Shelley, P. E. An assessment of automatic sewer flow samplers. USEPA, Office of Research and Monitoring, EPA-R2-73-261, June 1973.

Shelley, P. E. Sampling of wastewater. USEPA, Technology Transfer, June 1974.

Standard Methods for the Examination of Water and Waste Water. 14th ed. Washington, D.C.: American Public Health Association.

Thompson, R. G. Water-pollution instrumentation. *Chemical Engineering,* pp. 151–154, June 21, 1976.

REFERENCES

1. *Handbook for Monitoring Industrial Wastewater.* USEPA, Technology Transfer, August 1973.
2. Harris, J. P., and S. A. Kacmar. Flow monitoring techniques in sanitary sewers. *Deeds and Data,* Water Pollution Control Federation, July 1974.
3. Merrill, W. H., Jr. Conducting a water pollution survey. *Plant Engineering,* pp. 102–104, October 15, 1970.
4. Merrill, W. H., Jr. How to conduct an in-plant wastewater survey. *Plant Engineering,* pp. 41–43, January 7, 1971.
5. The Merck Index-An Encyclopedia of Chemicals and Drugs. 9th ed. Rahway, N.J.: Merck and Co., 1976, Mrs. M. Windholz, Editor.
6. Vernick, A. S. How to conduct a wastewater survey. *Plant Engineering,* pp. 85–87, July 7, 1977 and pp. 77–80, August 4, 1977.

7. Steffen, A. J. Understanding the publicly owned treatment works. UESPA, Technology Transfer Seminar on Pretreatment of Industrial Wastes, Philadelphia, Pennsylvania, May 24–25, 1978.
8. *Financial Assistance Programs for Pollution Prevention and Control Available Through the Federal Government.* USEPA, Office of Analysis and Evaluation, May 1977.
9. Urban Systems Research & Engineering, Inc. Financial strategies to alleviate the costs of compliance with pretreatment requirements. Environmental Protection Agency Technology Transfer Seminar on Pretreatment of Industrial Wastes, Philadelphia, Pennsylvania, May 24–25, 1978.
10. *Federal Guidelines—State and Local Pretreatment Programs.* USEPA, Publication no. EPA 430/9-76-017a, MCD-43, January 1977.

7

Case Histories of Industrial Waste Control Programs

INTRODUCTION

Selected case histories of municipal wastewater treatment plants' pretreatment programs are presented in this chapter. Some of the issues highlighted are area descriptions, water pollution control efforts, pretreatment program development, industrial compliance, and program effectiveness. While much attention has been focused on pretreatment in recent years and substantial information pertaining to pretreatment guidance, program implementation, and industrial roles exists, the extent of pretreatment programs' effectiveness in abating water pollution problems still needs further investigation. The case histories presented in this chapter include: the City of Grand Rapids, the Hampton Roads Sanitation District, and the Metropolitan Sanitary District of Greater Chicago. A review of the Buffalo Sewer Authority's (1) pretreatment program is included in Chapter 3.

CITY OF GRAND RAPIDS (2)

Area Description

The City of Grand Rapids, with a population of 190,000 within the city limits and 350,000 in the metropolitan area, is the largest city in Michigan's western lower peninsula. Industry is highly diversified in this area, although Grand Rapids is perhaps best known for its production of fine furniture and as having one of the largest concentrations of electroplating firms in the country.

The Grand River, in the heart of Grand Rapids, is an important recreational area for western Michigan. During the late 1950s and 1960s, severe

241

environmental deterioration of the Grand River caused by industrial discharges of cyanide and heavy metals resulted in periodic fish kills.

Water Pollution Control

During the mid-1960s, the need to reverse this damaging trend was recognized by the public. In January 1969, the Grand Rapids city commissioner enacted a comprehensive water pollution control ordinance (now referred to as the sewer use ordinance), establishing effluent limitations for cyanide and heavy metals, as well as other provisions. Prior to enactment, metal platers lobbied wholeheartedly against any limitations, arguing that the cost of pretreatment of wastes would push them out of business and force them to relocate. After much debate among city policy makers, the concern for the quality of the water and the general environment remained the focus, and the ordinance was adopted as city law.

Water pollution control is the responsibility of the city wastewater treatment plant, which in addition to serving Grand Rapids' wastewater disposal, also provides service to eleven cities and townships on a contractual basis. After expansion, the capacity of this activated sludge wastewater treatment plant is expected to be 90 mgd.

The sewer use ordinance and industrial sewer use regulations set forth the standards, rules, and regulations with which industrial users of the sewer systems must comply, as well as provisions for enforcement and management of the law. Of particular impact to industries were the prohibited discharge substances listed in the sewer use ordinance, as cited verbatim here:

1. Any effluent having a temperature higher than 140 degrees F.
2. Any effluent which contains more than 50 mg/l of animal fat, vegetable fat, oil or grease, or any combination thereof.
3. Any gasoline, benzene, naphtha, fuel oil or other inflammable or explosive liquid, solid or gas.
4. Any grease, oil or other substance that will become solid or viscous at temperatures 60 degrees Celsius and below after entering the System.
5. Any substance from the preparation, cooking and dispensing of food and from the handling, storage and sale of produce which has not been shredded to such a degree that all particles shall be carried freely under flow conditions normally prevailing in the public sanitary or combined sewer, with no particle larger than one-half inch in any dimension.
6. Any substance capable of causing obstruction to the flow in sewers or other interference with the proper operation of the sewage diposal system including but not limited to mineral oil, grease, ashes, cinders, sand, mud, plastics, wood, paunch manure, straw, shavings, metal, glass, rags, feathers, asphalt, tar and manure.

7. Any effluent pH lower than 6.0 or higher than 10.0 or having any other corrosive properties capable of causing damage or hazard to structures, equipment or personnel of the treatment works.
8. (a) Any effluent in excess of:
1.5 mg/l of Cadmium as Cd.
6 mg/l of Zinc as Zn.
2 mg/l of total Chromium as Cr.
1.5 mg/l of Copper as Cu.
1 mg/l of Cyanide as CN.
1.5 mg/l of Nickel as Ni.
0.2 mg/l of Phenol or derivative of Phenol.
(b) Any discharge of phosphorus, ammonia, nitrates, sugars or other nutrients or waste waters containing them which have an adverse effect on treatment processes or cause stimulation of growths of algae, weeds, and slimes which are or may become injurious to water supply, recreational use of water, fish, wildlife, and other aquatic life.
9. Any paints, oils, lacquers, thinners or solvents including any waste containing a toxic or deleterious substance which impair the Sewage Treatment process or constitute a hazard to employees working in the Sewage Disposal System.
10. Any noxious or malodorous gas or substance capable of creating a public nuisance.
11. Any effluent of such character or quantity that unusual attention or expense is required to handle such materials at the sewage treatment plant or to maintain the System.
12. Any discoloration such as, but not limited to, dyes, inks, and vegetable tanning solutions, or any unusual chemical oxygen demand, chlorides, sulfates or chlorine requirements in such quantities as to be deleterious and a hazard to the System and its employees.
13. Any radioactive wastes or isotopes of such half-life or concentration as may exceed limits established by applicable Local, State or Federal regulations.
14. Any effluent containing a five (5) day biochemical oxygen demand greater than 300 mg/l.
15. Any effluent containing suspended solids greater than 350 mg/l.
16. Any effluent containing phosphorus greater than 40 mg/l.
17. Any effluent having an average daily flow greater than 2% of the System's average daily flow.

Development of the Pretreatment Program

The effluent limitations stipulated in the sewer use ordinance and accompanying regulations required significant changes in industrial waste disposal practices and the investment of several million dollars by local industry in the design and construction of pretreatment facilities. As a result of the effluent limitations, the already strained relationship between the city and local industry became more tense. To alleviate this tension and to facilitate industrial compliance with the limitations, the city attempted to establish mutual cooperation through personal contact with representatives of affected companies in the area and by providing short-term variances to those companies.

Each company was granted a two-year variance to the ordinance which allowed them to exceed effluent limitations, provided that the progress made toward construction of a pretreatment facility was evident. The variance stipulated that industry must file a set of design plans with the city for its pretreatment system within six months and submit progress reports every six months thereafter. During the first six months of the program, city staff visited each of the companies to meet management personnel and to explain the details of the effluent limitations. It was explained that the city staff would be responsible for monitoring the industrial waste on a continuous basis and to recommend whatever legal action was necessary to achieve compliance.

Although industry had in most cases made a good-faith effort to meet the deadline, after the two-year period, it was recognized that effluent limitations were not yet achievable due to the delay in delivery of pretreatment equipment. A six-month extension in the variance was offered by the city as a shakedown period after which active enforcement would begin.

Following the two and one-half year variance period, industries were subject to penalties for effluent limitation violations. Court trials, where necessary, usually resulted in convictions and fines. Company fines usually consisted of the maximum city code penalty of $100 per violation, city costs of surveillance, and court costs. The heaviest fine levied against a single industry was almost $5,000, representing 41 violations of the ordinance plus city surveillance costs. In Grand Rapids, industry compliance with effluent limitations is ensured by a full-time surveillance program that is adequately staffed and equipped to conduct regular and routine effluent sampling in such a way that proper and reliable data results.

The news media have been instrumental in industrial compliance. Since the industrial progress reports made to the city commission were public information, they received a great deal of news media attention. At one point, seven electroplating industries sited as being lax in their effort to install pretreatment equipment were referred to as the "Dirty Seven." Public reaction against these companies encouraged them to achieve compliance levels and discouraged them from ever lagging behind again.

Collection and Disposal of Industrial Sludges

In order for the control of industrial wastes to be effective, proper collection and disposal of the residual materials resulting from the pretreatment processes is required. Sizable quantities of both liquid and solid metallic hydroxide sludge began to be generated in the Grand Rapids area once the pretreatment systems became operable. The liquid sludges (2–6% solids) were generated by companies that had no physical space for dewatering equipment or felt that liquid disposal was more economical. Other companies installed vacuum filters, centrifuges or other filtering devices to solidify their sludges to solid contents of 20–30%.

Since no specific state legislation or previous industrial sludge disposal experience existed, little attention was paid to the disposal practices. During the first five or six years of the industrial program, solid sludges were disposed with solid waste in sanitary landfills or placed in an approved, abandoned gravel pit site. Monitoring wells located near the gravel pit site showed a migration of heavy metals into the water table. As a result, the site was closed for metal hydroxide sludge disposal and state legislation was enacted to establish standards for land disposal of industrial sludges. Separate sites for disposal of industrial wastes are required, and the standards for these sites dictate sufficient clay thickness (depths of 25 to 30 feet) to prevent any migration in either a vertical or horizontal direction.

A lack of licensed sites in the Grand Rapids area for the disposal of sludges has forced many companies to contract to have their wastes hauled long distances for disposal, in most cases to Illinois or Indiana. As a result, the cost of transporting and disposing of sludge has drastically increased. To add to this dilemma, when a local site meets all the criteria of the state licensing regulations, it is often blocked by public reaction and pressure in the township where it is proposed. Pressure from neighboring states where the sludges are being deposited has also developed. Efforts are being directed at creating enough local disposal sites to prevent industry from disposing its waste into the public sewer and reverting the City of Grand Rapids to its pre-water pollution control condition.

Program Effectiveness

Significant reductions in the heavy metal concentrations found in the Grand Rapids plant's influent and effluent have been evident since the adoption of the pretreatment ordinance. Comparative influent–effluent total metal concentrations for the years 1968 through 1978 are illustrated in Figure 7-1. While influent levels have dropped from 12–13 mg/l to about 2 mg/l, effluent levels have dropped from 9–10 mg/l to about 1 mg/l, representing 87% and 92% reductions, respectively. Reductions of chromium (90% and 96%), copper (89% and 93%), nickel (87% and 89%), and zinc (79% and 85%) are illustrated in Figures 7-2 through 7-5. The erratic changes in zinc levels are thought to be connected with air pollution control requirements which, in 1970, forced brass foundries to install scrubbers to remove zinc oxide from their air stacks. This waste material was then discharged into the sanitary sewer system, causing zinc concentrations to soar to previous levels. Attention was then directed to methods of achieving pretreatment that would reverse this trend, and by 1973 levels of zinc in the sewage system dropped sharply. Total cyanide, as illustrated in Figure 7-6, shows excellent concentration reductions of 93% and 96% for influent and effluent, respectively.

Relative reductions between influent and effluent show that even though lower concentrations were present in the effluent prior to pretreatment operations, higher reductions were obtained after pretreatment. This was observed for all metals and cyanide. It appears that wastewater treatment plants are capable of removing or treating low levels of metals but that efficiency decreases with increasing influent concentration. This nonlinear

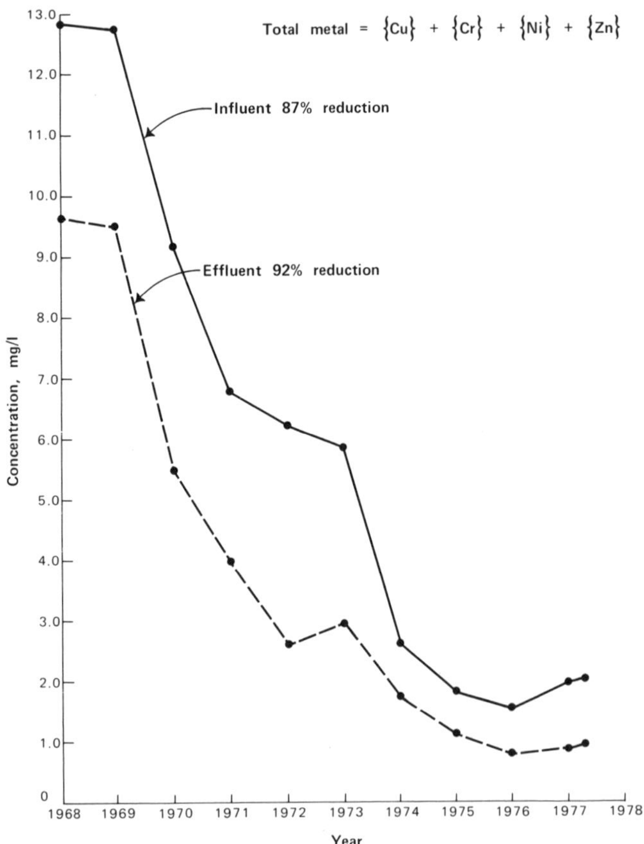

Figure 7-1 / Total metal in sewage, City of Grand Rapids.

removal efficiency is exemplified through comparison of the removal of each metal at the higher influent concentrations before pretreatment to the lower levels after pretreatment. Percentage removals are illustrated in Table 7-1.

Slug discharges, common to any batch operation, can have adverse effects on a wastewater treatment plant's operations, particularly on its biological systems. These short but very concentrated discharges, typically impair treatment efficiency for periods ranging from minutes to days. Prior to the pretreatment enactment slugs were common; such occurrences are now rare.

In the wastewater treatment process, much of the influent heavy metal is removed with the sludge. The reductions in heavy metals in the sludge average 66%, resulting in sludge that contains only about one-third the concentration of heavy metals as the preordinance sludge. This cleaner sludge enhances the feasibility of land application in Grand Rapids as a viable resource recovery practice that aids in reducing costs of incineration.

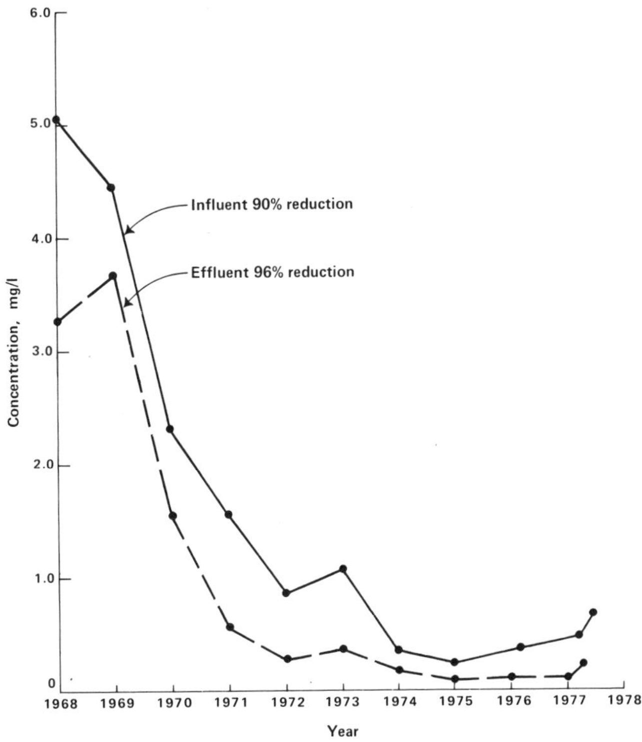

Figure 7-2 / Total chromium in sewage, City of Grand Rapids.

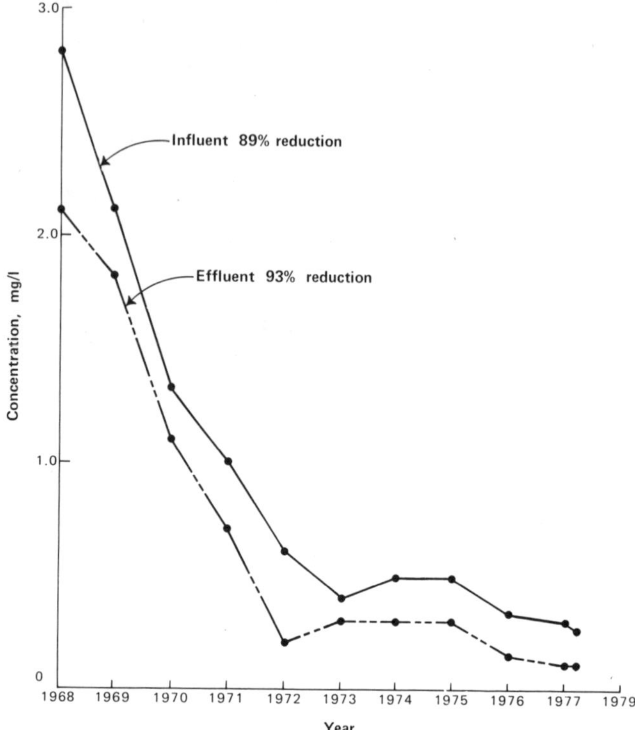

Figure 7-3 / Total copper in sewage, City of Grand Rapids.

Perhaps the most illustrative example of the pretreatment program effectiveness is in the revival of the Grand River as a recreational resource. A river which, because of industrial contamination, had, from the late 1950's to early 1970s, been the victim of periodic fish kills was by 1974 supporting a large population of trout and salmon. This was due largely to the reduction in levels of toxic wastes being introduced to and alternating in the river. Cyanide levels in the river have dropped from 120 to 4 ppb, while dissolved chromium levels have dropped from 340 to 15 ppb (3). Canoeing enthusiasts began promoting and mapping the river. The construction and improvement of boat ramps was initiated. More parkland was purchased by the city and developed into recreational areas along the banks of the river. Through its pretreatment program, the City of Grand Rapids transformed the deteriorating Grand River into a well-respected natural area and a highly prized fishing and recreational site.

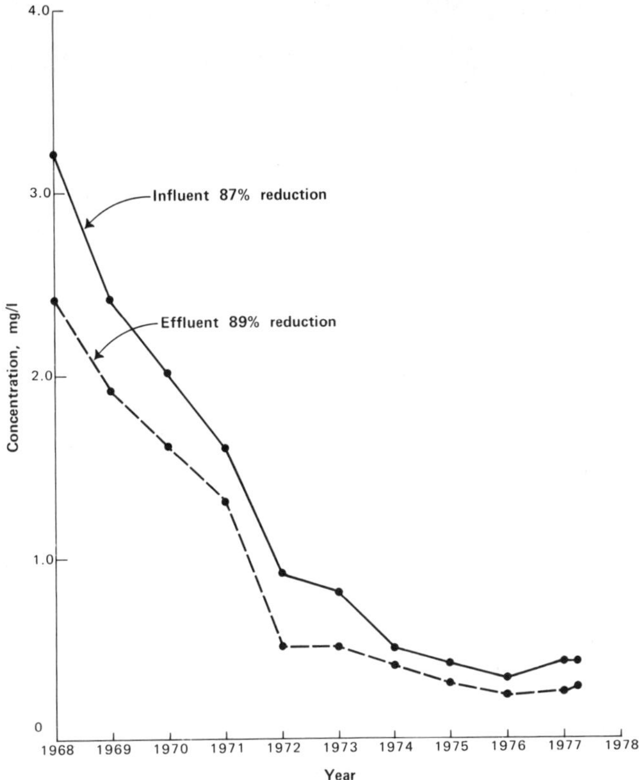

Figure 7-4 / Total nickel in sewage, City of Grand Rapids.

HAMPTON ROADS SANITATION DISTRICT (4,5)

Area Description

The Hampton Roads Sanitation District (HRSD) is a political subdivision of the commonwealth of Virginia charged with the responsibility for waste collection and treatment within its jurisdiction. The district's boundaries currently encompass approximately 1,685 square miles of the southeastern Virginia Tidewater region. The district is unique in that it serves an area surrounding Hampton Roads, which is the largest natural harbor in the world.

The Tidewater is a very heavily militarized area. More than half of all industrial waste generated within the district comes from large military installations. This presents some difficulty in locating sources of problem waste, since these facilities are themselves similar in size to small cities. Other

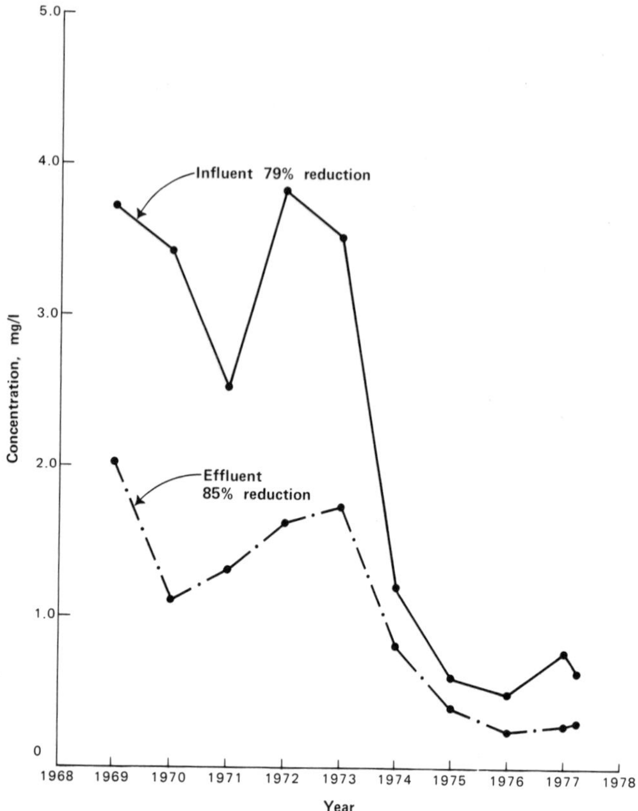

Figure 7-5 / Total zinc in sewage, City of Grand Rapids.

sources of industrial wastes include: light and heavy manufacturing, food processing, and associated service industries. Wastes generated by these industrial and commercial sources are treated at the district's nine treatment plants. These plants treat, in total, approximately 130 million gallons per day. The treated effluent is presently discharged into the Elizabeth River, James River, Chesapeake Bay, and Hampton Roads Harbor. Discharging into the Atlantic Ocean is proposed.

Water Pollution Control

Unlike the case of City of Grand Rapids, noticeable environmental deterioration did not provide the impetus for the development of a pretreatment program. On the contrary, significant problems related to the quality of receiving waters was not documented. Industrial waste control, which began

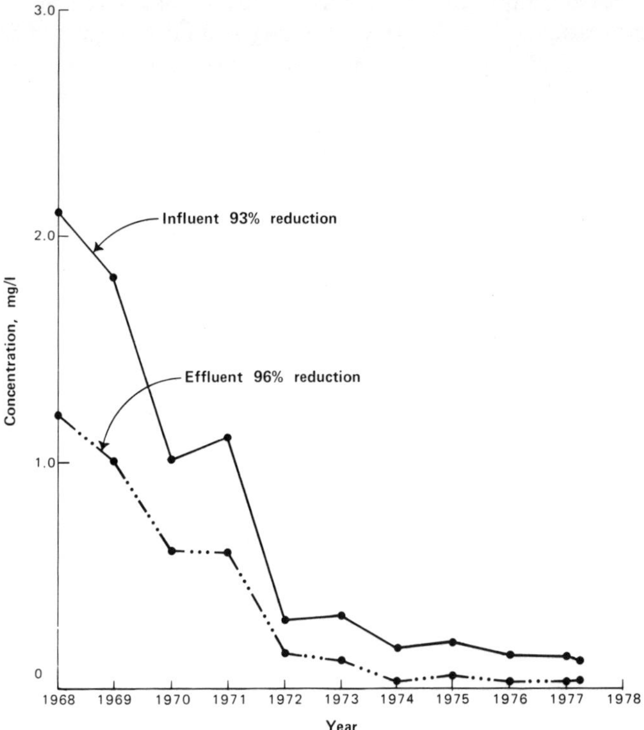

Figure 7-6 / Total cyanide in sewage, City of Grand Rapids.

Table 7-1 / City of Grand Rapids, removal efficiencies for various constituent influent concentrations before and after pretreatment.

| | Before | | After | |
| | Influent concentration mg/l | Percentage removal | Influent concentration mg/l | Percentage removal |
Constituent				
Cyanide	2.1	43	0.14	71
Chromium	5.1	35	0.49	73
Copper	2.8	25	0.30	63
Nickel	3.2	25	0.42	38
Zinc	3.7	46	0.78	63

in 1972 was prompted by the realization that industrial dischargers were upsetting treatment plant operations and that sludges with high heavy metal concentrations which were undesirable for land application were being produced. In order to meet the NPDES standards and to operate the treatment plants effectively, the industrial waste control program was initiated.

Although EPA final pretreatment limitations for industrial subcategories were not available during program development, existing industrial waste problems necessitated the prompt enforcement of HRSD limitations. During the development of HRSD's specific pollutant limitations, two key factors were taken into consideration. The first factor considered was protection of HRSD treatment structures and processes. HRSD scientists carefully reviewed proposed limits (particularly heavy metals) for their effect on the quality of sludges generated by the district, taking into account ultimate disposal. Although incineration is now the principal method of sludge disposal used by the district, land application is planned to replace incineration in the future. In addition, effects of the proposed limits were examined for various processes, including activated sludge and anaerobic digestion, using sampling data where possible. Effects of incompatible pollutant discharges to receiving waters from HRSD treatment facilities were also studied to ensure that state and federal water quality requirements were satisfied.

The setting of local standards that would be compatible with EPA pretreatment limitations presented considerable difficulty, since EPA had not issued any standards. Local standards were set based on known problem areas and proposed EPA standards, where these could be obtained. Local standards much less stringent than EPA limits, could result in construction of inadequate pretreatment facilities and wasted capital investments, since EPA limits must be enforced when promulgated. Conversely, local standards much more stringent than EPA limits and not based on specific treatment objectives (e.g., land disposal of sludge) could result in "treatment for treatment's sake," wasted energy, and duplication of effort. It was HRSD's intent to develop and enforce local standards compatible with EPA "final" limits. It is believed that pretreatment systems designed and built to meet HRSD limits could be the "guts" for whatever system may be necessary to meet EPA limits, for example, by addition of polishing filters to an existing chemical precipitation system for heavy metals removal.

Industrial Waste Division

HRSD's industrial waste division is responsible for regulating industrial and commercial discharges through application of the district's industrial wastewater discharge regulations. These regulations include general and specific effluent limitations and a discharge permit program. Compliance is effected through the use of an extensive monitoring program in conjunction with industrial self-monitoring. HRSD conducts annual workweek wastewater monitoring surveys utilizing high-speed automatic samplers interfaced with flow monitoring equipment. In addition, all major contributing industries and other industries known to discharge toxic pollutants are grab sampled and spot checked monthly at unannounced times. These same industries are required to perform varying degrees of self-monitoring using methods which are acceptable to HRSD.

Industrial discharges to HRSD's facilities are issued discharge permits which outline applicable effluent limitations, monitoring requirements, metering requirements, and appropriate compliance schedules. Of the 78 industries which discharge wastes requiring pretreatment to meet HRSD limitations, 61 have already achieved compliance. Of the 17 not in compliance, six are in the construction phase of necessary pretreatment systems and the remaining 11 have been, or are being, placed on compliance schedules. One industry (a truck manufacturing plant) is currently investing 4.5 million dollars to achieve compliance.

HRSD's industrial waste division currently implements a surcharge program generating in excess of 1 million dollars in revenue annually. The surcharge is an additional charge to industrial and commercial users which have high-strength wastes (BOD and/or TSS concentrations greater than 250 mg/l). The surcharge serves a twofold purpose:

1. High-strength users are required to pay their fair share of treatment costs rather than being supplemented by residential customers generating normal domestic wastes;
2. Industries are given a financial incentive to maintain pretreatment objectives.

Many industries have substantially reduced or eliminated the surcharge by instituting simple housekeeping practices, thus saving themselves money and providing additional treatment capacity at the POTW.

An important part of industrial waste control at the POTW is the ability to locate and hold liable sources of unusual discharges which cause treatment difficulties. HRSD has such capabilities in the form of an "industrial waste alert system." This system provides for 24-hours-per-day response in investigating and terminating discharges of unusual wastes which damage HRSD treatment structures or interfere with treatment. This activity has resulted in the collection of considerable sums in damage from industrial sources of unusual wastes.

Program Effectiveness

Since implementation of the pretreatment program, noticeable improvements in treatment plant operations and in the quality of sludge generated have resulted. Records of plant upsets due to highly concentrated industrial discharges have been reduced to virtually zero. Sludges of seven of nine wastewater treatment facilities are now suitable for land application. Sludges of the two remaining plants are close to approaching suitable quality and it is expected that their utlization for land systems will soon be feasible. HRSD's industrial waste control program is designed so that the best and most cost-effective wastewater treatment services can be provided to all of its customers.

THE METROPOLITAN SANITARY DISTRICT OF GREATER CHICAGO (6)

Area Description

The Metropolitan Sanitary District of Greater Chicago (MSDGC) services the 865 square-mile area in northeastern Illinois. The MSDGC operates six sewage treatment plants that treat, in aggregate, approximately 1.5 billion gallons of domestic and industrial wastewater daily. The MSDGC is also responsible for monitoring all discharge into 72 miles of navigable inland waterways, over 200 miles of small rivers and streams, and 36 miles of Lake Michigan in-shore waters. An estimated 11,000 industries, of which approximately 6,000 are considered wet (i.e., discharging some process wastewater and/or cooling water) are within the jurisdiction of the MSDGC.

The Sewage and Waste Control Ordinance

The MSDGC sewage and waste control ordinance, adopted in 1969, includes concentration limits for 14 contaminants and 9 limiting conditions for discharge of industrial waste to the sanitary sewer system, and includes concentration limits for 26 contaminants and 7 limiting conditions for direct waterway discharges. In 1978, the MSDGC amended its sewage and waste control ordinance in order to be able to trace industrial sludges from source to ultimate disposal. The ordinance requires pretreatment facilities which generate sludges which contain certain contaminants in concentrations greater that those prescribed in the ordinance to report to the MSDGC the quantity and composition of the sludges and to identify the scavenger or hauler and the ultimate disposal location.

Problems relating to the disposal of pretreated sludges have already developed. A number of incidents have occurred in which landfill operators have refused to accept certain waste loads from scavengers. Until the ordinance amendments required that the quantity and composition of sludges be reported, landfill operators were under the impression that scavengers were disposing garbage instead of industrial sludge. The landfill operators' reasons for refusing to accept the waste were twofold: first, the discharger and scavenger were paying for the disposal of garbage and not industrial sludge; second, in many cases, the landfill operators were unknowingly accepting hazardous wastes. Since the sanitary landfill operators have become aware of the origin and character of wastes, the cost of disposal of these wastes has risen substantially.

Industrial Case Histories

Compliance with the sewage and waste control ordinance for many industries involves only process changes and improved housekeeping; for some industries, however, compliance requires the installation of pretreatment facilities. In most cases, pretreatment operations produce a sludge which is difficult and expensive to dispose of, and they often require large amounts of energy to operate. Discussions of three brief case histories of industries in the MSDGC are presented to illustrate typical pretreatment systems, sludges produced, and energy consumed. These three industrial processes include: leather tanning, captive metal plating, and industrial laundering.

Leather Tanning

The pretreatment system operated by the industry consists of waste equalization, automatic pH control, chemical flocculation, clarification, centrifuging of the sludge that results from the clarification process, and the recycling of approximately 30% of its process waters.

In the treatment of approximately 0.5 mgd, 30,000–60,000 cubic yards of sludge are generated annually. The present costs for disposal of this sludge is $0.47 per cubic yard. These costs have increased more than 23% in the past three years, and are expected to increase at an even higher rate during the next three years. The sludge generated from this company contains organic chemicals, fats, oils or greases, copper, chromium, iron, lead, and mercury, and is classified as a hazardous waste.

By recycling 30% of the treated industrial wastewater, the company has been able to reduce the amount of water used per unit or production from 172 gallons to 117 gallons, a reduction of 25%.

Captive Metal Plating

The pretreatment operation of this industry accomplishes one of the major goals of EPA, which is the reuse of process waters. Although reuse has been shown to be more costly than discharge to the public sewer system, the company adopted this concept to express their environmental consciousness.

Three evaporative recovery systems for the recycling of effluent from their cadmium, chromium, and nickel plating lines are in operation. These systems do not produce sludge. This waste treatment system also utilizes automatic pH control, and a 26,000 gallon equalization tank that, in turn, discharges to a dissolved air flotation system for the removal of oils and greases from various metal-cleaning operations. The dissolved air flotation system presently produces approximately 120,000 gallons of sludge per year.

This sludge costs $0.09 per gallon to dispose of, and the cost of disposal has increased 60% over the past three years. The total cost for operating this pretreatment system is $10,800 for sludge removal, $22,500 for various chemicals, and approximately $70,000 per year for energy. The system also requires three full-time people for operation, at an estimated additional cost of approximately $100,000 per year, including overhead. The total estimated cost for operating this industrial waste pretreatment system, less replacement cost, is $203,300 per year, about one-third of which is assigned to energy.

Industrial Laundering

This pretreatment system, rated at 100,000 gallons per day, consists of equalization, chemical flocculation, and dissolved air flotation. Vacuum filters are used to dewater the sludge developed by the dissolved air flotation process. In 1978, this system treated 19.6 million gallons of wastewater and produced 1,135 cubic yards of sludge. The cost for sludge removal and disposal, which is usually in a sanitary landfill, is $5 per cubic yard. The overall cost of treatment in 1977 was $1.27 per 1,000 gallons of wastewater treated.

REFERENCES

1. *Industrial Waste and Pretreatment in the Buffalo Municipal System.* USEPA, Environmental Protection Technology Series, EPA-600/2-77-018, 1977.
2. Biener, J. A., and W. H. Bouma. Case history—City of Grand Rapids, Michigan, program of industrial waste control. *Pretreatment of Industrial Wastes—Joint Municipal and Industrial Seminar.* USEPA, Technology Transfer, Seminar Handout, 1978.
3. Heritage, J. Industrial pretreatment. *EPA Journal,* vol. 5, no. 7, p. 17, 1979.
4. Industrial waste control program: An overview. Hampton Roads Sanitation District. (Unpublished.)
5. Aydlett, G. M., chief of industrial wastes division, Hampton Roads Sanitation District. Personal communication, 1979.
6. Whitebloom, S. W., R. Lanyon, and C. Lue-Hing. Impact on sludge production and energy consumption for industrial pretreatment systems resulting from USEPA pretreatment regulations. The Metropolitan Sanitary District of Greater Chicago, May 1979.

Index